A MODERN DOG'S LIFE

Praise for *A Modern Dog's Life*

"*A Modern Dog's Life* combines sensible information with charming wit; this is an entertaining guide for new and veteran dog owners alike."
—KEN FOSTER, author of *The Dogs Who Found Me*

"Science, experience and common sense—Paul McGreevy brings all three to this amazing new book. Your dog will want you to read it."
—MARK EVANS, Chief Veterinary Adviser, RSPCA UK

"A milestone in understanding dogs . . . Perhaps even the Rosetta stone in translating dog behaviour . . . Dr. McGreevy is a world-recognized expert in dog behavior who just happens to have written a book that is as easy to read as a quality novel . . . If you want to have more fun with your dog, you need this book. If you just want to know more about dogs than your friends, you need this book."
—DON BURKE, founder, *Burke's Backyard*

"I've been waiting for a book like this for years. A clear and logical account of dog behavior and training . . . both scientific and entertaining."
—DR. BIDDA JONES, Chief Scientist, RSPCA Australia

"I would encourage all dog owners to read this book, to understand their dog's view of our human lifestyle and to improve the human–canine relationship."
—DR. JOANNE RIGHETTI, Chair of the Delta Society Australia

"Dr. McGreevy's book is an essential text for any true dog lover's library. It gives the new dog owner the chance to start on the right foot, the seasoned dog owner the knowledge to improve life for themselves as well as their pet, and even the experienced dog trainer the opportunity to see many things better through a dog's eyes."
—ROB ZAMMIT, DVM

THE EXPERIMENT

BECAUSE EVERY BOOK IS A TEST OF NEW IDEAS

A MODERN DOG'S LIFE

How to Do the Best for Your Dog

Paul McGreevy, PhD, MRCVS

THE EXPERIMENT
NEW YORK

A MODERN DOG'S LIFE: *How to Do the Best for Your Dog*
Copyright © Paul McGreevy 2009, 2010

The Experiment, LLC
260 Fifth Avenue
New York, NY 10001–6425
www.theexperimentpublishing.com

First U.S. edition published by arrangement with University Press of New South Wales, Australia.

Many of the designations used by manufacturers and sellers to distinguish their products are claimed as trademarks. Where those designations appear in this book and The Experiment was aware of a trademark claim, the designations have been capitalized.

Library of Congress Control Number: 2009940034
ISBN 978-1-61519-018-8

Cover design by Michael Fusco | michaelfuscodesign.com
Cover photograph by Karen Mork
Author photograph by Pierre Malou
Back cover photograph by Paul McGreevy
Design by Josephine Pajor-Markus

Manufactured in the United States of America

First printing May 2010

10 9 8 7 6 5 4 3 2 1

Contents

For my late father, "Elvis,"
and all my beloved hound-dogs,
past and present.

A MODERN DOG'S LIFE

1

For the Love of Dog

Dogs and humans have evolved alongside one another over a long period of time, but all is not well in the Land of Dog. We breed dogs with negligible emphasis on temperament, even though problem behaviors are the main trigger for euthanasia in young dogs. We want dogs that are devoted to us but somehow expect them to cope when left alone. We persistently frustrate our canine companions by ignoring what they truly value. This book is about dogs' needs and how we can improve our understanding of dogs and how best to look after them in the twenty-first century. Drawing on the latest research and my expertise as a veterinary behaviorist who has spent a lifetime with dogs, my aim is to suggest a new approach to owning a dog. I hope to explain why dogs thrive on three key things: fun, exercise and training. Most importantly, I offer fresh ideas about how we, as dog owners, can help our dogs access these goodies.

Salman Rushdie described dogs as the "loving, half-comprehending, half-mystified aliens who live within our homes." *A Modern Dog's Life* looks at aspects of our behavior that are particularly mystifying to dogs and establishes why dogs may never comprehend some of our characteristics and tendencies. It also examines features of dog management that many owners struggle to get right—and sets out some blunt home truths about the realities of keeping a dog. Ultimately, *A Modern Dog's Life* is for anyone who wants to understand more and therefore demystify their dog. Its aim is to help you to become a better dog-watcher, team player, caregiver, companion and life-coach by knowing when and how to intervene.

This is not a book about the charm of dogs or the many ways of caring for them. There are hundreds of such books already out there. Instead, my premise is that owning a dog takes time and thought and is not always a pleasure. Despite figures pushed out annually by pet food manufacturers as they insist that pets are good for our health, we all know that dogs can also cause tremendous distress to humans around them, and not only to their owners. This book asks why dogs can be distressing and why they get distressed. It offers solutions to some common doggy dilemmas but does not shy away from the fact that many dogs lead less than ideal lives. In a sense then, this book is for those who strive to do the best *for* their dogs rather than those who need to get the best *out* of their dogs.

My aim with this book is to deliver insights and challenges that prompt you to reflect on your own dog's behavior. All the dogs you have shared time with offer examples of the concepts I describe. When exploring the unwelcome consequences of our actions on the welfare of dogs, I promise not to use the trite and inadequate remark: *How would you like it?* This is not useful because our chief challenge is to think like dogs rather than expect them to have the same sensitivities we have. My pledge is to avoid interpreting dog behaviors in human terms. Any statement suggesting that dogs are almost human is, for many dog enthusiasts, nothing short of an insult. In return, I encourage you to use my reflections to improve the lot of the dogs you know now or are yet to meet. This book gives dogs the benefit of the doubt (and of the latest research) when it comes to their feelings, but never assumes that they have human intelligence. Dogs have canine intelligence—for them, a far more useful attribute.

As we gather more information on dogs and their behavior, we begin to realize how much there is still to discover. Humans owe dogs a great deal, and vice versa. We have coevolved, exploiting one another to various degrees. Indeed, we continue to do so in novel ways that I note throughout this book.

What is "natural behavior" for a dog?

Dog keeping may be as old as hunting, grunting and cave painting, but studying domestic dogs in family homes is a complex business. Each dog's behavior and motivation may seem clear enough, but they usually reflect human differences. One family may lavish attention on their dog, while another virtually ignores theirs. One person within a family may be a great trainer, while another, within the same household, may be inconsistent or incompetent. If we want to understand dog behavior as clearly as possible, the most helpful observations come from populations of free-ranging dogs living in the wild, uncontaminated by direct contact with humans. No collars, no leashes, no bowls, no beds, no fences. Such dogs come from the same stock as our domesticated dogs but live separate from humans. Completely unpolluted data can be very difficult to obtain. Although free-ranging dogs tend to stay away from disruptive and dangerous human activity, they often are still affected by people. Even dogs living on a landfill can be influenced by the humans who deliver the rubbish, while those hiding in remote forests and undeveloped land can be disturbed by human activity at the boundaries of their territory. Dogs considered feral may have been dumped as pups and so are products of the human–dog interaction.

Traditionally, we have tended to regard the wolf as the perfect model of what dogs are like without human interference. To an extent, this is entirely valid, since we believe dogs evolved from wolves. The domestic dog is a subspecies of its ancestor, the gray wolf. Indeed, at times in the chapters that follow, I will refer to the gray wolf as Uncle Wolf as a nickname for the archetypal lupine forebear. And to save time, when offering examples of free-ranging or feral dog behavior, I shall refer to feral dogs as Feral Cheryl. The critical DNA sequences of the domestic dog differ from those of the gray wolf by only 0.2 percent. This means the two are very closely related and explains why they can interbreed. By contrast, the difference between the gray wolf and its closest wild relative, the coyote, is around 4 percent.

Given that dogs and wolves are virtually indistinguishable genetically, the enormous variation in body size and shape in the dog is truly remarkable. For example, whereas an adult wolf usually weighs around 100 pounds, an adult dog can weigh between 2½ and 200 pounds (obesity can send this upper limit even higher—but more on that in chapter 6, Sex, Disease and Aging). The breadth of behavioral differences that accompany these variations is also extraordinary.

Although the wolf is a popular model for canine behavior, Australian dingoes are probably a better one. Sadly, they are under threat in their pure state, because there are now so few that have not been crossed with modern breeds. However, their behavior is more that of the unfettered dog than any wolf's will ever be. Behaviorally, dingoes respond to their pack members in ways that are rarely evident in wolf packs. For example, adult dingoes play with one another far more than adult wolves; they vocalize more and are generally more flexible in their responses to strangers. In these ways they are typical of dogs. These behavioral differences are just the tip of the iceberg, since the assumption that all dogs behave the same way is as flawed as the notion that they all look the same. Breeds, after all, were originally the physical manifestation of the human desire to distill particular behavioral traits,

CHEW ON THIS

During the process of domestication and the development of breeds, the skull characteristics of dogs have changed considerably. Skull length in adult dogs can vary between 2¾ and 11 inches, whereas in adult wolves it is around 12 inches. Unsurprisingly, the organs within the skull have also changed. For example, the brain-to-body weight ratio of domestic dogs is one-third of that in wolves, so a 100 pound wolf has a brain three times heavier than a 100 pound dog. Although it is clearly flawed to suppose that a dog is a dog is a dog, with this statistic in mind, I've become interested in how the entire nervous system, including the brain, may differ from one breed to another. Clearly, differences in the nervous system have profound implications for the differences in behavior of different breeds.

often accompanied by recognizable shapes, color and coat lengths that can act as markers for those behaviors.

As we explore the science of dog behavior we must accept that much of what we think we know is still only speculation. It would surprise most dog owners to discover that academic animal behavior journals report many more studies on bees than on dogs. Why? The average person spends a great deal more time with dogs than with bees, so surely we need to know more about dog packs than bee swarms? Purists might argue that bees are more interesting than dogs to serious students of animal behavior (ethologists) because their behavior is less the product of human interference in the form of genetic selection and husbandry. It is almost as if familiarity has bred ignorance. Happily, I can report that domesticated species have recently become the focus of vigorous scientific study as the field of applied ethology emerges to help solve behavioral problems. The bad news is that, among the domestic species being studied, dogs are bringing up the rear because they are regarded as less important than more commercially productive species, such as pigs, cattle and chickens. Perhaps this is a small price to pay for dogs not being regarded as a food source in the Western world (although, with the advent of fusion cuisine, Chow Chow and chips may not be that far away).

A note of caution

With any research effort, it always pays to ask: Who funds the study? Usually, costs are justified if there is a human benefit. Wealthy countries that use dogs in military service devote significant dollars to researching their behavior. Pet food manufacturers often fund studies that explore the benefits of pet ownership and ways in which pet ownership can be made easier. Guide dog associations may support studies that make dogs generally healthier or more successful in training. All of the above have human benefits: War dogs keep us safe from terrorism, pet dogs keep us happy, and guide dogs keep partially sighted people from becoming partially flattened.

Given that most research benefits humans, what about studies that benefit dogs? Much of the work some stakeholders would *not* wish to be associated with is funded by animal-welfare charities. Conspicuously little has been done in this domain, but the strides that have been made recently should be celebrated. That is part of what I hope to achieve with *A Modern Dog's Life*. I also hope to excite you with the prospect of a rosier future for dogdom.

Dogs respond to model dogs in intriguing ways that tell us a great deal about what is most relevant to them. So it is worth considering whether the skills they use when bonding with humans are entirely novel.

2

The Challenges for the Modern Dog

It's easy to forget that dogs have only recently begun to adapt to life in the modern world—a world filled with man-made design and technology. Although this world hasn't been around all that long, humans can rationalize about what's going on in it. On the other hand, the sights, sounds and smells of the twenty-first century might sometimes overwhelm dogs. Wheels, fire, electricity and chemistry are examples of the mechanisms we use to explain "magic" in the modern world. Our dogs experience the outcomes of these inventions without knowing the relationship between cause and effect.

Negotiating a shifting physical world

Consider the physical enclosures we have built around dogs. Solid walls were not part of a free-ranging dog's world—other forces kept pups in the vicinity of the den while the pack went hunting. Modern boundaries and surfaces, such as polished floors, electric fences and escalators can even be hazardous. Stairs, especially those with open spaces between steps, take a bit of getting used to. Then there are elevators that must feel, to a dog, like earth tremors as they come to rest. And how weird it must be for them to go into a room (the elevator) and exit through the same door to an entirely different set of stimuli.

And, of course, there's the challenge of doors themselves: some that open with a gentle nosing, others that slam shut with the wind. There are sliding doors, roll-down shutter doors, glass doors that dogs can see through and screen doors that dogs can see and even smell through. Then there are all those door handles that seem to be the trigger humans touch to make the door change position. Handles come in a raft of different sizes and shapes with locking mechanisms that can be correctly unpicked only by some devilishly deft dogs. The problem-solving these talented safe-crackers have processed is an outstanding example of trial-and-error learning and a tribute to their perseverance. We will explore their adaptive learning in some depth later in the book.

In the same way, cars are boxes that dogs enter only to find themselves emerging elsewhere. Of course, these are no ordinary boxes. If they look through the windows of these special, noisy boxes, dogs see changes: other dogs flashing past without moving their legs, dogs that disappear with or without being barked at (although many dogs seem convinced that barking helps to get rid of them). And when the noisy boxes come to standstill, the dogs who have traveled in them often score a walk in a new territory. How exciting! The joy of the noisy box is enormous for some dogs. No wonder they cock their legs against them.

So cars are extremely significant for many dogs. Dogs can distinguish between one car and the next and recognize the engine noises of different vehicles. Why? This is because certain cars are associated with certain humans. Familiar humans use familiar cars. Intriguingly, dogs can attach importance to the cars that important humans *depart in*, rather than *emerge from*. It is almost as if they can make an association between the noise the car makes once the significant human is inside it. The alternative is that they make the association retrospectively after the significant human has emerged from it. But this would require them to log all sounds of all cars just in case a significant human emerged from one of them, a taxing and time-wasting occupation. This skill in dogs is fascinating, since it implies that evolution has helped dogs accomplish the task of associating a novel noise with a disappearing member of the pack. Very puzzling, indeed! I warn you, we're still speculating our way through much of dog behavior. So, for many of these puzzles, your educated guess is as good as mine.

The glass used in windows of cars and houses offers an outstanding example of the mysteries of a modern dog's life. Modern materials defy canine reasoning. Dogs cannot know that this barrier through which they can see but cannot smell is a baked, silicon-based fluid that permits the passage of light particles. When puppies first encounter a glass panel they simply learn that its lack of permeability is nonnegotiable and that anything they cannot smell through is something they cannot walk through.

Who is that dog in the mirror?

The visual challenges of the modern world do not stop there. Consider mirrors. When a pup sees its reflection for the first time, what follows is usually amusing for most human observers. The learning curve this puppy is on is steepened by the fact that puppies generally find it difficult to identify objects until they reach visual maturity at four months of age. For a pup, a mirror placed on the floor reliably reveals a puppy galloping headlong towards him, staring, head-cocking and play-bowing. As the weeks and months of adolescence sweep by, the puppy trapped behind the mirror ages and become less interested in and interesting for the observer, until ultimately the two ignore each other almost entirely. Mirrors can help members of other species, such as horses and some birds, cope with isolation, but there is no evidence that they can spare dogs the misery of separation-related distress (discussed further in chapter 5, [Networking Among Dogs]). As yet we don't know whether this is a hint at self-awareness. The lack of response to the image in the mirror could mean that learning has helped to label the image as irrelevant. This passivity contrasts with reports from primate studies in which observing animals interact with their reflection and, most compelling of all, use the mirrors to remove spots of liquid paper that have been applied to their faces without their knowledge (during general anesthesia).

Television: What's all the fuss about?

If mirrors are confusing for dogs, television programs are probably an utter mystery, bringing as they do a nonstop cascade of moving images and sounds that humans gather round to sit and stare at. In the natural state, dogs in a group never arrange themselves around an object and stare at it. The closest equivalent is the way they might surround a prey item, and then the standing and staring gives way very rapidly to grabbing and tearing. Televisions don't smell like prey and they don't move like prey, so do dogs look at humans paying homage to the colored cabinet in the corner and wonder what all the fuss is about?

Interestingly, if they react to a TV set at all, dogs are much more likely to respond to the sound it makes than the light it projects. This suggests that the images are difficult for dogs to resolve. The speed with which their brains process images (the so-called flicker fusion speed) differs from ours and explains why few dogs seem to respond to supposedly relevant images, such as those of other dogs. Some dogs react equally to a ball as to a sheep moving across the screen and sometimes rather charmingly inspect behind the television set to find these items once they've disappeared from view. At best it appears that dogs see quadrupeds rather than dogs. Most dogs that respond to animals on a TV screen do so equally to horses as to cattle, and to aardvarks as to antelopes.

Humans and their ever-changing ways

As if houses, cars and TVs were not enough of a challenge, humans keep changing. Their shape, color and smell simply cannot be relied upon. Sometimes they wear dark blotches over their eyes (sunglasses) that reliably prevent a dog from seeing what its owner is looking at. Clothes can change the appearance of even the most familiar human, a hat can change the outline of an owner dramatically and, to add to a dog's confusion, humans carry things (sometimes as large as ladders and barrels) that defy doggy understanding. After all, no dog would ever be able to carry these in its mouth let alone its paws, so why would it have evolved to expect this morphing magic?

Meanwhile, the modern olfactory world humans impose on dogs is filled with unnaturally strong odors, including perfumes, aftershaves, air fresheners and household cleaning agents, the olfactory equivalent of the blaring white noise that comes from what baby boomers call boom boxes. Plainly, we cannot know precisely what the modern dog makes of all these novelties and how he copes with all these challenges In chapter 17 we will examine innovative ways in which technologies can help us to meet our dogs' behavioral needs. Meanwhile, let's focus on what we actually know about the dog's senses.

How dogs perceive the world

SENSE OF SMELL

This is a dog's predominant sense, allowing it to discriminate odor molecules among complex mixtures of odors. The dog has about 220 million scent receptors in the nose, whereas humans have only 5 million. Anatomical differences aside, the power of the dog's sense of smell has been tested at 10,000 to 100,000 times stronger than that of humans, an order of magnitude equivalent to one second in 317 centuries. It has been said that you never have to motivate a dog to use its nose. To me, this means sniffing is what all dogs do all of the time. To sniff is to be canine.

Detector dogs are becoming increasingly popular as wars on terror and drugs continue apace. Temperature, humidity, wind and age of the trail as well as the strength of the odor all affect the success rate. Scenting the broken vegetation associated with footprints helps dogs track humans rather than the waft of specific odors that the target person might leave in his wake. Dogs are even able to pick up the scent of disease, with published results indicating their impressive ability to sniff out the distinctive chemicals produced by cancerous cells in skin, urine (from bladder cancer cells) and even on breath (from sufferers of breast cancer with a 88 percent accuracy and of lung cancer with 99 percent accuracy).

The *vomeronasal* organ, originally but incorrectly thought to be exclusive to nonhuman animals, is an additional component of the sense of smell. Its key job is to detect pheromones, the secreted chemicals thought to facilitate the mother–infant bond and mediate territorial and sexual behavior. In dogs, the organ is located in the roof of the mouth just behind the upper incisors. Dogs put the vomeronasal organ to work by flicking their tongue in and out of the mouth, almost as if drinking. Urine is the main vehicle for pheromones, which is why dogs spend so much time finding and anointing optimal marking points and take so much care sniffing those that have been visited by others. Pheromones do not appear in feces but are smeared on it in a fine strip from the anal sac (gland) as it passes through the anal sphincter. Anal sac secretions contain pheromones that differ from one group of animals to the next. These differences suggest that individual dogs may detect age and genetic differences when assessing others' feces and under-tail odors. However, all dogs are not equal when it comes to scent detection.

A dog's success in tracking the source of an odor or detecting a target scent also depends on environmental conditions. Wind direction can affect the concentration of odor molecules while warmer temperatures increase the rate of panting, a response that inhibits a dog's ability to draw in sufficient air to get a really clear "picture" of the smells around him. This has been neatly demonstrated in studies showing that sniffer dogs are less effective on hot days.

SENSE OF SIGHT

A dog's vision is generally inferior to that of humans, but it can see color, static shapes and considerable detail with its central visual field. Having said that, dogs are very sensitive to moving objects, and there is compelling evidence that some can see a human arm waving up to almost a mile away. Dogs are very sensitive to sudden or unusual movement, a useful asset in guide dogs, retrievers and hunting dogs. The panoramic field of vision is 250-270°, but binocular vision varies greatly in different breeds according to how far apart their eyes are set in their heads (for example, Pekinese and bull terriers have about 85° binocular vision, greyhounds about 75°, while man has about 140°).

CHEW ON THIS

A dog's peripheral vision depends on its skull shape. We have studied an arrangement of cells in the retina to explore this aspect of vision. A concentrated band of cells across the equator of the retina, the visual streak, is necessary for peripheral vision, but strangely it has disappeared in short-skulled breeds such as the pug. In the species we have studied to date, only the dog and the horse, there is a direct correlation between nose length and the concentration of critical ganglion cells in the visual streak— long nose, long streak; no nose, no streak. We have yet to establish why this is so.

Although it was previously thought that dogs are color blind, recent studies have shown that under bright light dogs can detect wavelengths within the blue and yellow portion of the light spectrum and are therefore dichromatic. However, they can't distinguish reds and oranges as they have only a few of the cones sensitive to the red/orange wavelengths. The visual color spectrum of dogs can be seen in two forms: violet and blue-violet, which is seen as blue and greenish-yellow; and yellow or red, which is seen as yellow. Therefore, dogs are red-green color blind but are better at differentiating between shades of gray than humans are.

The predominance of rod receptors in their retinae allows dogs to see much better than humans do at night. Their absolute threshold for the detection of light is about threefold lower than humans, allowing them to be three times as capable of detecting low light intensities. The *tapetum lucidum*, located behind the retina, maximizes light within the eye and so assists the dog's night vision. Its reflective cells form the greenish-yellow layer we see in their eyes, classically when they look into the beam of car headlights at night. Intriguingly, some dogs, most notably chocolate Labrador retrievers and some merle-colored dogs, lack this specialized coating. We don't yet know how this affects their ability to see, but it could mean that they would be less comfortable traveling around at night. They might be more likely to bump into things or to be more wary, especially in strange surroundings, because they have less ability to discern shapes.

SENSE OF HEARING

Dogs have a highly developed sense of hearing and can hear high notes the human ear cannot detect. Children can detect notes up to frequencies of about 20 KHz, adults rather less, while dogs are known to be able to hear notes up to 35 KHz, and it's been suggested that their limit may be as high as 100 KHz. Small wonder that they can hear keys rattling, the postman's bike and bacon being unwrapped so well. Such acute hearing is probably most useful for capturing small prey, such as mice and bats, which emit high-frequency sounds to communicate. While sounds can be detected up to 40 KHz, there is no evidence that dogs can communicate with each other at such high frequencies (by ultrasound).

Intriguingly, the RSCPA in the United Kingdom has called for the banning of loud fireworks, claiming that they are a form of cruelty to dogs. The staggering numbers of dogs that end up in dog pounds after a night of firework blasts seems to support the view that dogs cannot cope with the noise. By voting with their feet and leaving their homes, they are telling us that the relative security of the den, which they have evolved to value so highly, has been eclipsed by the terrific roar of an unpredictable beast that finds them even when they hide in the very tightest spots. The therapeutic effects of gentle exposure to thunder recordings are well recognized and have prompted studies to establish dogs' preferences for piped music as a form of environmental enrichment in rescue kennels. When compared with human conversation, heavy metal and pop, classical music caused dogs to rest more and bark less.

Just as dogs' ears vary in size, length, shape and hairiness, they also differ in their ability to detect sound. We'll learn about the advantages or disadvantages of the various ear shapes in chapter 14, which looks closely at breed differences. Some dogs are born deaf, especially those with reduced pigment in their coats. Because auditory nerve cells and *melanocyte* (pigment cells) develop from the same part of the embryo, defects in the pigmentation process are often linked to defects in the auditory pathway. Merle dogs (such as Wally and Neville, whose photographs appear on the cover of this book) are the best example of hypopigmentation; merles should never be mated with other merles because this increases the chances of their offspring being deaf.

It's a great tribute to their trainers that congenitally deaf dogs have succeeded at the highest level in obedience trials. Once any dog with normal hearing has learned the basic responses to a single signal, it should be trained to respond to both verbal and visual commands. This will allow it to relate to visual cues alone if it happens to lose its hearing later in life.

SENSE OF TOUCH

It's difficult for us to imagine what the world feels like to a dog. A dog's skin is to an extent buffered from the immediate effects of wind, water and even solid surfaces by fur, but this buffering can be removed by clipping, sometimes with striking results. Many owners report wholesale changes in their dog's behavior after swimming, but even more so after clipping: tripled joie de vivre with a cherry on top is the most common report. Wriggling, squirming and play-bowing, freshly clipped dogs race around like hellhounds, spin like tops, roll like pins and break as many rules as possible. They really let their hair down. Why? Perhaps it is simply release from the burden of a heavy coat that lets them exercise more readily when they have been clipped (Fellhounds used in racing after scent-trails in the north of England are often clipped to aid heat loss). Or maybe they are reminded of how their skin used to feel when they were puppies . . . or maybe they even *know* they look different and are somehow amused. Whatever the reason, they often bounce around and cavort like lunatics.

The non-furry bits of a dog seem less sensitive than many parts of the human body surface. The leather covering the pads and the nose tip is perfectly designed to be tough, thick and resilient because it is in an exposed area. The nose leather prevents injuries being incurred during fighting, play-fighting and the intense sniffing in soil and dead leaves that is called for when digging. Meanwhile, the leather on footpads is subject to wear and tear whenever the dog is not lying down. It has to withstand not only abrasion and penetration but also heat and cold. The downside for the dog is that these leathery coverings probably deny the dog an ability to feel surfaces as we do. But then again, what would the dog do with such information? We need tactile fingertips for activities such as grooming the surface of our skin; dogs do not.

One area that seems more tactile than its equivalent in humans is the muzzle. The *vibrissae* (whiskers) of the dog are mobile, and each has a blood-filled sac at its base to amplify movement. Next time you are relaxing with a dog you know, try brushing past its vibrissae ever so gently with your fingers. The slightest contact causes reflex wrinkling of the lip, a response that must help in fights by exposing the teeth whenever a combatant touches the side of the face. We can't say for sure how else the dog's whiskers help it detect the world around its nose. Maybe they are indispensable for digging and generally rooting around.

CHEW ON THIS

Captive seals in aquatic shows use their whiskers to help them balance balls on their noses. This is an excellent demonstration of how whiskers may help detect prey that, being at the tip of the nose, has fallen out of sight but not out of reach. The world's most elaborate and exquisitely sensitive whiskers are found in walruses; they use them to pick up the telltale signs of mollusks on the seabed. It is possible that dogs' whiskers have a similar purpose.

SENSE OF TASTE

Food passes down the gullet of the average dog with such speed that it spends very little time in contact with the tongue and its array of taste buds. Dogs sample food with experimental lickings if they don't recognize familiar odors on approach. So-called lapdogs seem particularly good at this, with Maltese having notoriously capricious appetites. They do not wolf food down as if there were no tomorrow. On the contrary, they seem confident that if they hold out for something tastier they will be rewarded. Many owners of such princely (and princessy) couch potatoes train their dogs to behave this way. The fact is, you get the behaviors you train . . . and the dog you deserve.

If capriciousness can be learned, then so too can genuine caution. As we will see in chapter 11 (The Artful Dodgers) food-aversion learning allows dogs to use a "suck-it-and-see" policy when encountering new foods. Dogs on medication involving daily or more frequent dosing with tablets stealthily placed in lavish snacks are particularly cautious with novel foods. They are especially wary when unmedicated food is rammed into their mouths without them having to work for it. One system of heading off such suspicion is to give the lavish snacks more frequently at times when dosing is not required and as part of regular training. The secret here is to require the dog to offer a trained response before feeding any snack without and, most important, with medication.

While for most dogs taste is not an important part of eating, it certainly plays a critical role in sexual surveillance by male dogs. We are all familiar with the lip-curling flehmen response shown by horses and bulls (routinely and erroneously labeled a laugh by tabloid journalists). The canine equivalent involves physical pushing of volatile odors from urine towards the vomeronasal organ with a very particular sort of licking action. As part of courtship, dogs will lick bitches' ears, lips and genitals. This latter target is explored for detection of physiological readiness to mate, whereas the ears and lips are touched as a means of testing the bitch's tolerance or her behavioral readiness. Clearly, some of the things dogs choose to lick make many humans gag. For example, to the horror of some human onlookers, bitches clear up the urine and feces from their nest-bound pups with the same relish older dogs retain for eating cat feces. Dogs of either sex also seem to enjoy licking their own body excretions, but again it isn't clear whether this waste management points to good or bad taste. One minor source of comfort for owners is that when dogs do this they cannot get infections from their own excretions.

How dogs communicate— with each other and with others

Now that we have some ideas about what dogs can detect with their senses, let's look at how they use these senses to communicate with one another and with members of other species.

SCENT-MARKING

The roles of the sniffer and the sniffed are profoundly important to dogs when they meet one another. The sniffed generally has his ears pinned back. He may be required to stand still when the sniffer has finished studying his rear end and has moved to the back of the neck—the axis of canine communication. He is vulnerable and has to focus his surveillance on any early warning of trouble. He can spring to one side with hair trigger responses. Sometimes he picks up false alarms and may try to fling himself to one side to avoid being approached from behind, almost as if he is trying to keep his secrets. This cautiousness reflects a number of attributes ranging from simple agility and bounciness (a young dog can usually outpace an older, heavier model) to experience. Some dogs have a very low threshold for warning signals, especially if they've had bad experiences in the past or have simply been denied opportunities to socialize as youngsters. Also, high-ranking dogs will move out of the sniffed role and into the role of sniffer with considerable speed. The challenge for two dogs on first meeting has a lot to do with diplomacy about exchanging these roles. Moving too quickly can elicit stand-over tactics, neck pressing and even biting.

A dog examining a stick will first smell it, perhaps for traces of saliva from a previous canine user, or perhaps a trace of urine that is sometimes left as if to make the item less inviting to other dogs.

CHEW ON THIS

Evidence from studies of wolves suggests that urine marking of emptied food caches may denote them as empty and improve the efficiency of foraging.

Marking is a way of being seen or, more commonly, smelled. Such is the power of dogs' olfaction that, in the strictest sense, marking in the form of depositing odors happens whenever dogs or their body fluids make contact with a solid surface. Regardless of whether or not it is intentional, marking assumes many forms, from scraping the feet after defecation, to nesting, to rolling. Urine marking is clearly more intentional and is very important to both male and female dogs when (or these days more commonly, if) they reach sexual maturity and attain high rank. Some dog owners feel it cruel to stop their dogs from reading what they regard as the olfactory newspaper when they stop at every lamp post. You can always spot these dogs (and their owners) because they are both conspicuously under-exercised. These dogs hold onto the contents of their bladders on home soil regardless of the discomfort of a full bladder. This price is modest for the opportunity to mark. The cost dogs are prepared to pay for the opportunity to mark while off their home patch is further reflected in the neck pain and throat occlusion they undergo while straining against collars to remain near a prime marking spot until their work there is done.

Walks mean smells. Remember that novel walks may be arousing to dogs because in the free-ranging state new territory would be most unusual. Feral Cheryl and her brigade do not go to new areas every day or get into cars and hit the beach for an hour. Instead, they generally stay put in their own territory. And why wouldn't they? They know their territory well because they have explored it. Knowledge of the territory allows them to exploit all the resources it has to offer and provides them with the best chances of survival and reproduction. In other words, it enhances biological fitness.

CHEW ON THIS

One of the recent studies from my laboratory highlighted the importance of marking for dogs. It showed that dogs use their feces in marking novel areas sometimes to the extent that they can induce diarrhea. This behavior is especially common in entire (that is, non-castrated) males, followed by neutered males and then females. There were also differences in the way the dogs deposited their feces, with entire males being more likely to mark on vertical surfaces such as trees. The testosterone-driven motivation to mark a territory is worth close consideration. In free-ranging (rather than domestic) dogs, an entire male encounters new territories when he has left the family group. The chances are he may be straying dangerously close to another (potentially very hostile) group's territory. So why would he risk alerting that group to his presence and furthermore risk the dehydration that comes from diarrhea?

The need to be acknowledged by other dogs would seem to be the prevailing goal here. As social animals, dogs need to attract the company of other dogs so much that they are prepared to pay a hefty price. In a bid to advertise their presence and availability, males need to find prospective mates by pinning their smelly posters to anything sufficiently pole-like. It would be interesting to explore the responses of females after they have sniffed feces of both neutered and entire males.

When we exercise dogs in unfamiliar areas, they love the challenge of all those new odors, uncharted terrain and the wealth of opportunities. Dogs are clearly joyful as they explore all this with us, their social groups, but when alone their response may be very different. In the free-ranging state, the move from a home territory is usually a sequel to disaster or expulsion or both. So the concept of enjoying novel terrain may at first seem unlikely. To an extent, the dog's opportunistic nature may help counter any trepidation. As we'll see many times in this book, opportunism is what makes dogs so very successful in the ever-changing human world that is the home of most companion animals. Behavioral flexibility is what makes dogs so adaptable but, equally, it's also what allows us to imagine they are coping when they probably are not.

Scraping the feet after defecation and, to a lesser extent, urination is a fascinating response. Although at first glance it may appear similar to a burying action, it's not thought to have a function in covering waste. First of all, if that were the function, there seem to be very few dogs that are any good at it, and also, it is not a technique used for burying really important things, such as bones. Second, the dogs that scrape have generally just spent a deal of time planting their feces in critical spots (under trees, beside lamp-posts and in clumps of grasses) so why in Dog's name would they ever want to cover them up?

The more likely explanation for scraping relates to leaving marks. The marks in question are twofold: visual and olfactory. The visual signal comes from the scrape marks in the ground, while the olfactory components take the form of freshly turned soil that must ooze odors that alert passing dogs to the possibility of recent activity in general and to the musky scent of dog foot in particular. Dog foot odor is strangely charming to most dog owners and predictably disgusting to everyone else. For me it smells of freshly rained-on earth and field mushrooms, but for one friend of mine, it is sometimes evocative of warm corn chips. Next time your dog is falling asleep beside you, gently draw a foot towards you and inhale so that you can make up your own mind.

POSTURE

The posture of a dog can change considerably from a soft, fawning play-bow to a cringe of submission to a stiffened tonus of arousal. Play-bowing is a signal that is becoming recognized as being very basic in that it is clearly understood by almost all dogs. It is given almost exclusively to forward-facing recipients and may be especially important in adult dogs, when playful challenges may have the worst consequences if misinterpreted. With practice, humans can mimic it convincingly. They can also mimic canine stalking, but this should be attempted only with care. Stalking can be part of both play and serious predatory aggression. When a dog rolls over it can usually, but not always, defuse an apparently aggressive encounter with another dog, especially if the roller smells of puppy. Exposure of the soft underbelly is regarded as a clear sign of submission. Spraying with urine may emphasize this deference.

If a low-slung body means deference, haughty postures speak of considerable status. The higher a dog raises its head, the more easily it can impose itself on others with neck-pressing, pawing, mounting and shoulder barging. The stiff-legged stance is also a common response to unfathomed threats. Dogs, especially males, often follow this stance with high-trotting, an extravagant elevation of the knees. This form of display is used to approach an animal or object of considerable interest. These key postures aside, visual communication in dogs is largely a matter of heads and tails.

HEADS AND TAILS

Heads can be hairy, smooth or covered in skin-folds. All of these affect a dog's ability to signal with its eyes, ears and mouth. The eyes can deliver stares, signals that dare the recipient to take one more step, but only if they are not covered in a veil of hair. To name but a few variants on the typical upstanding ear of Uncle Wolf, the ears of dogs can be pendant, rose-shaped or like those of a bat. All of these are more or less mobile and therefore vary in their eloquence when interacting with other dogs. Similarly, lips may be soft and droopy or bearded and baggy, again varying in their ability to communicate warnings and intentions. When you see a Russian wolfhound greeting a pug, you can't help being amazed that dogs recognize each other as dogs at all, let alone get along so well.

As we have successfully fiddled with the shape of the proto-dog to give our breeds incredibly diverse appearances, we have unwittingly also removed varying amounts of signaling capacity. So far, only a few breeds have been assessed for their ability to signal but, in general, the more stylized the breed, the fewer signals it can issue. The French bulldog, with its reasonably immobile ears, bunched-up face and absent tail, is a great example. The evidence that golden retrievers are actually able to issue more signals than German shepherd dogs might come as a surprise to many readers, since the shepherd is so wolf-like, but we shouldn't assume that because a dog looks like a wolf it can signal like one. It appears that the present-day German shepherd has ancestors that were far less lupine than they currently appear, and so the wolf-like features we see today have, in fact, been bred back into these dogs without necessarily being functional.

One breed that is yet to be assessed is the Old English sheepdog. It is easy to imagine a fair number of signals that are compromised by its shagginess, including the ability to raise the hackles, deliver a fixed stare or bare the teeth. Add to this the docked or naturally bobbed tail and you can see why one might predict that dogs of this breed struggle to communicate with each other, let alone other dogs. This implies that they may well be more likely to fight than other dogs because they are misread. In the absence of any evidence that they get into more fights, it seems possible that they have been bred with a temperament that compensates for their signaling deficits. So maybe it has always been important when breeding Old English sheepdogs to select, perhaps even unconsciously, for tolerance (that is, to select for a high threshold to aggression).

Yawning may well be a way of increasing circulation to the head or, more specifically, the brain, but it also has a function in communication. It is believed that yawning may have calming properties for both yawning and observing dogs. It may even have a role in social facilitation by helping to defuse tense situations. Yawning is often accompanied by lip-licking, body-shaking and sneezing in dogs confronted with a challenge that can range from being left outside a shop to being stared at by a stranger. Time spent watching dogs under challenging circumstances, such as when waiting their turn in a veterinary clinic, can be very enlightening. Although the context of these signals gives tremendous insight into their roles, we can't fully explain their meaning yet.

Incredibly, rigorous evidence of dogs using their tails to communicate is very scarce. This has played into the hands of those who support docking, since they can claim that the tail is surplus to a dog's requirements. People have docked the tails and cropped the ears of dogs for numerous historical reasons, but in modern times none of these is a good enough excuse. Those who insist on removing bits from their pups will always attempt to justify their actions, probably because their guilt makes them acknowledge how much dogs that have tails actually use them. They are the flat-earthers of the dog world, and, as the tides of veterinary and public opinion turn against them, they would do well to stop fighting the current.

Whether the tail is used to balance, signal or spread odors, or all of the above and more, the chances are that dogs without them do learn to compensate. Many of us see docked dogs wagging their entire rear end, and there are even occasional stories of docked dogs damaging their necks from too enthusiastic full-body wags. This should remind us that the repercussions of breeding bizarre shapes and cutting bits off puppies are not fully understood. We tinker with these things for our gratification, but it is our dogs that pay the price. That said, dogs have a remarkable ability to communicate adequately with stylized members of their own species despite all of our tinkering with canine body shape. In this regard, they certainly surpass horses, some of which freak completely when a familiar herd-mate turns up wearing a new weather-proof blanket. This ability may contribute to the success of dogs in the human domain, because it may contribute to their specialized skills in reading human-given communication gestures (especially pointing and, to a lesser extent, head turning and glancing).

CHOICE CUTS

- Cars, elevators, man-made boundaries and surfaces all offer challenges to the modern dog.

- The dog's sense of smell may be as much as 100,000 times stronger than that of humans, so understanding the olfactory world that dogs live in offers a significant challenge to us.

- Dogs have a highly developed sense of hearing and can hear high notes the human ear cannot detect.

- Dogs use their ears, tails, lips and eyes to communicate in ways that eclipse human nonverbal communication.

- A dog's ability to signal visually and pick up visual signals from other dogs depends on its breed.

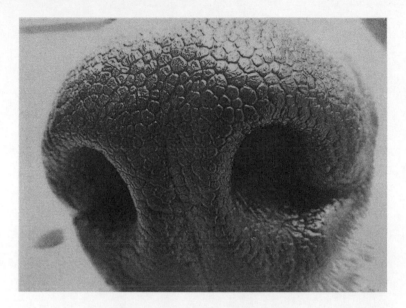

We will never know what it is like to live in a world of potent odors but can only wonder at the sensitivity of the dog's nose, not least to pheromones. It is likely that the modern dog finds air fresheners, detergents, perfumes and aftershaves challenging, if not aversive.

Foot-pads are important sources of odors for dogs. Even humans can discern differences between the pad scent of one dog and another.

3
What Dogs Value

The most important question for the thoughtful dog owner is: *What do dogs value and why?* The answer will help us assess our dog's behavioral needs. Instead of merely accepting that dogs like food, we should ask: What is it about food that makes them work for it and fight for it? Instead of simply marveling at a sniffer dog's approach to detective work, we should consider why its tail is wagging. Instead of just knowing that dogs enjoy exercise, we should explore what an exciting walk can offer a dog and how a bland walk or, worse still, no exercise at all can frustrate it. All of these questions help us to avoid judging dogs by our own standards. To understand what a dog values is to take a walk in its paws.

Consumer demand theory

Animal welfare science has blossomed over the past two decades. Among other studies, it measures the needs of domestic and captive animals and helps us to rank the importance of the resources we offer them. One of the most common principles in this work is called

consumer demand theory: the harder you are willing to work for
something, the more you must value it. So, when assessing an animal's demand for a given commodity, counting the number of
responses it makes to get the resource shows how much work the
animal is prepared to invest. This relies on the animal viewing the
resource as a reward. In the early stages the animal needs to make
only a few responses to obtain a particular reward. The number of
responses required is increased by some fixed amount after each
reward until the break point—when the animal stops responding. At
this point, the animal is saying, in effect, the price is too high. For
some resources an animal may be prepared to pay any price. In the
language of economic theory, such demands are *inelastic*.

Of course, such preference measures can only judge the immediate appeal of some object, situation or substance, and may not always
reflect wise choices. Like people, animals sometimes choose things
that in the long run are harmful, such as sweet foods and addictive
drugs. In experiments on addictions in animals, it's been found that
the break point can be very high, so it would be a mistake to assume
that animals always know what's good for them. For example, dogs
would work long and hard for the fat and sugars in chocolate, but
these ingredients can cause diabetes and obesity. Even more important in the short term, the theobromine content of chocolate can
cause death by over-stimulating the heart.

The good stuff: food, drink, fun, company and comfort

Although they have yet to be "costed out," food and drink, fun, company and comfort seem to be key resources for dogs of all ages. By
watching dogs as they respond to certain things, we get to know
what they regard as the good stuff. It might be the smell of cheese in
the kitchen or a glimpse of the park through the car window. It

might be the sound of Scooby Snack packaging or the leash being rattled. It could be the sound of the front door opening or of a ball being thrown. What dogs do with these resources, how they rank them and how they work to access and defend them, gives us fascinating insight into what it means to be canine. The lesson we must learn as dog watchers and handlers is when to use the most valued resources as bargaining chips in trying to shape the behaviors we want. Some of you may recoil from such a calculating approach to a human–animal bond, but to dismiss this strategy is to pass over some extremely valuable tools and run the risk instead of becoming your dog's dutiful servant. If you are one of those people who deplores requiring animals to work for a living and would rather just give them endless pleasure and liberty to do as they wish, consider getting a cat instead of a dog.

Learn from your dog and be an opportunist, too

This chapter explains what it is like to be a dog in a human social group. Dogs are opportunists, and when you share your world with a dog you'd better know how to be an opportunist, too. Not only will you understand what dogs are looking for in life, you will understand how they optimize their environment and how you, as their life-coach, can become the source of their very best opportunities.

A bowlful of training opportunities

Dogs don't spend long enough tasting food to reliably detect the difference between one snack and another (Uncle Wolf would wonder why manufacturers bother to push new flavors of dog food, chews, snacks, training rewards, dental "bones" and pig body-parts). While he made do with rabbit every day of the month, his descendants can expect a constantly changing menu with a semi-digestible garnish and a side order of aspic. We imagine that a dog's need for variety and treats is the same as ours, and pet food marketers applaud our naïveté as we hurl ourselves happily into their trap. Indeed, dogs have evolved to investigate new types of food and new ways of sourcing it, but they can manage very well on one well-designed diet. Their bowels thrive on consistency, and they never notice a lack of chef's specials, seasonal menus or *plats du jour*.

However, we can capitalize on the opportunistic nature of dogs by using new and different foods in training. When we want to let the dog know that it has done really well, we pay out a jackpot. Making the jackpot especially novel and tasty increases the value of the reward and makes the dog work harder in the future (much more of this in chapter 10 (The Dogs of Opportunity) and chapter 12 (Fine-Tuning).

Dogs are generally well behaved before they are fed. For example, many sit beautifully, with the canine version of a winning smile, as their food is delivered. Unfortunately, very few owners realize that when they feed their dog its evening meal, they are holding a bowlful of training opportunities. As the "holders of the resources," we should consider mimicking the sniffer-dog trainers who insist that their dogs earn each bit of their keep (their daily ration) by offering trained responses. We can usually estimate the value of a food item to a dog in terms of its energy content or novelty. Dogs generally prefer a lump of cheese to a lump of carrot and a slice of roast beef to a can of economy dog food. Their desire to obtain preferred resources is reflected in the work they are willing to perform to get them—dogs are at their most attentive at feeding times. We also observe dogs offering almost their entire repertoire of trained responses to trigger us to give a highly fancied reward. By the same token, dogs will pay high costs to defend valuable resources, costs that may even include risk of injury.

CHEW ON THIS

In many cases, dogs' attitude to food provides a convincing argument for them being unable to project well into the future. In contrast, rats in so-called cafeteria-style feeding trials that offer all-you-can-eat (*ad libitum*) feeding for set periods will alter their intake at one meal in anticipation of the size of the next. Dogs do not seem to have this skill. As extreme opportunists that evolved to eat as a competitive social activity, dogs are obliged to start eating at every opportunity and to keep wolfing the food down until it's all gone. Clearly, some breeds are more challenged than others when it comes to learning moderation. Labradors and beagles are always at the back of the class drooling over their lunchboxes. Members of other breeds, such as Maltese, are notoriously fussy, especially if their owners fold easily under emotional pressure and have plenty of time on their hands to stare at uneaten food.

Bones and chewing

For a dog, chewing is an activity as essential to its being as licking its genitals, pulling its lip back when scratching its neck or scooting its anus along the ground to relieve irritated anal sacs. Chewing is a means of discovery. Chew something for long enough and you can swallow it to see if it will cause illness. Chewing is also a source of comfort. Only by chewing things can a dog meet its needs for oral satisfaction when left alone. It's worth considering how many wild dog pups would ever be left by their pack without a barrage of chewable items from sticks to bits of prey (hide, hooves and bones) to play with and nibble on. Chewing makes life better, especially for youngsters, and even more especially for youngsters who are teething.

Dogs place considerable value on bones, even those with no apparent calorific content (such as bones without any meat or gristle attached, or artificial bones, such as Nylabones™). In addition, dogs commonly chew bones at times of arousal and after eating a meal. This suggests that chewing bones might help digestion and possibly even reduce stress. So, it's advisable to give dogs plenty of opportunities to chew. However, two main stumbling blocks arise—from other dogs and pet food manufacturers. Other dogs regularly compete (fight) for access to bones, even when plenty are available, while pet food manufacturers claim that dogs' diets are complete and that further supplementation is unnecessary. Strictly speaking, commercial diets are adequate nutritionally, but their physical structure fails to meet a dog's need to chew for a prolonged period. Who knows why pet food manufacturers emphasize the complete nature of their diets so strongly? Perhaps they'd rather owners did not experiment with other foods.

Veterinarians may appreciate the enjoyment dogs get from chewing bones, but they're also aware of the dangers of foreign bodies and so are reluctant to suggest feeding bones, since these may cause a future intestinal obstruction or penetration. Nobody wants a dog to chew greasy, gristly bones indoors, so I recommend a nylon equivalent as a means of meeting your dog's fundamental need to chew. Helping the dog learn what to chew and where is the only challenge that remains. My own dogs have all been quick to learn that the inside (synthetic) bone does not go outside—and so does not get covered in mud—and the outside (real) bones stay outside so that the carpet does not get covered in blood.

Toys and possessions

Recent studies have shown that dogs issue a distinctive growl when forced to guard their food and that, when played back, recordings of this characteristic sound reliably deter most dogs from approaching an unattended bone. Bones, bits of hoof, rags are all chewable, but so are chair legs, shoes and clothes off the washing line. These items can therefore be as precious to a dog as the toys we supply. Balls, sticks, rope tugs and so on are simply the items we designate as appropriate dog playthings. We add value to them by putting energy into them—throwing them, seizing them, possessing them. The extent to which we can add value in this way is limited by our enthusiasm but, more importantly, by our perceived leadership. The leader determines which is the preferred toy. Running after a pup that has picked up an Italian leather slipper may seem the most obvious thing to do, but it usually sends the pup at least three potentially troublesome messages: You have my attention; you have got a valuable item in your mouth; I am your follower. A more astute response is to spring to life, cavort and gambol to the nearest appropriate item and play with it as ostentatiously as modesty will permit. The item could be anything vaguely plausible as a new toy for the pup . . . a paper tissue, a newspaper, even a pair of knickers can be sacrificed to save the slipper. When the pup drops the slipper, reward it with a game with the more appropriate item.

If we can add value to an item by playing with it, dogs can do the same. When a dog sees another dog entering its domain it is likely to run up to it and say hello. Next it confirms whether it recognizes the incoming traffic as a previous play partner. If it does, then it usually races to its favorite toy to reduce the chance of it being misappropriated. Some behavior counselors encourage this as a way of giving dogs something "constructive" to do when human guests arrive. It is an interesting approach, since it can invite play and certainly fills a mouth that might otherwise do something as potentially unacceptable as lick the visitors. Of course, the downside is that the toy may be a stuffed animal that visiting children see as cuddle-worthy. If they immediately reach for the article, they may be bitten. Indeed, if they go to pick up the item some time later, the dog may rush to defend it.

The same urgent rush for resources is sometimes seen in parks and other areas shared by dogs. Some may rush for sticks as play-things and tug-of-war ropes; they race to trees to mark them, and they run towards other dogs to investigate their potential as play-mates, sexual partners or challengers. Why the rush? Because they are social animals that have evolved to compete with their group for all the good things in life, and once they have possession, they take pride in it. Pride can explain why dogs can appear quite ostentatious with a prized possession. Instead of skulking around with it and enjoying it covertly, they will flag up their tails and parade the trophy for all to see.

Water is a reward

Drink, or more specifically, water, is easily overlooked as a reward. This is perhaps because it is so often readily available, since all pet-care books rightly, but rather relentlessly, insist that dogs should always have plenty of fresh drinking water. When a dog really needs a drink he will pull out all the stops to access water, even if it means slurping a parking lot puddle. This is a clear signal as to how valuable water is to a thirsty animal. Water can be used to lure feral pigs into a one-way funnel of fencing around a waterhole, and horses can be trained to enter trailers for access to water. So dogs could certainly be trained to undertake very challenging and even dangerous activities on the promise of a drink. But would it be right to train a dog in this way? We'll discuss the manipulation of motivation in a later section, but for now please just consider the price a dog will pay for its various nutritional needs, including water. Plainly, the price will tend to vary according to transient demands. The need for water will vary with ambient temperatures, humidity and exercise, among many things.

The internal motivation to drink depends on cues such as a dry mouth, concentrated blood (after a salty meal) and reduced blood volume. These are the physiological stimuli, but there are other prompts that aren't so well understood. Social facilitation is one that's worth considering. Just as the sight of a model chicken pecking at the ground can trigger foraging behavior in a well-fed observer bird, so the sight of a dog drinking can stimulate an observing dog to join in. The other prompt is the return of the pack. Many owners report that when they return home, their dogs greet them and then trot to the water bowl and drink with great gusto. It may be that these dogs are trying to capitalize on a limited resource and so rush to the bowl in case their owners have the same goal in mind. Alternatively, the distress of being left alone may be sufficient to cause both apathy and an adrenaline response that dries the mouth. The apathy means that they do not bother to drink, and the adrenaline leaves them with a dry mouth. When their owner returns, they are roused enough to register that they are thirsty, a sense that may be especially strong if they have been snoring.

Dogs in transit provide a good example of dogs with an unusual need for water. Air-conditioning often dehydrates dogs on long journeys, so this is when fresh water really should be available at all times. However, because of danger of serious spillage, it rarely is. Next time you stop on a long road trip to give your dog a drink, notice how attentive he is as you ease the lid off his bottle of water. He is highly motivated. Becoming aware of your dog's behavioral needs identifies you as a good owner but also empowers you to be an excellent trainer.

In terms of their response to novel foods, dogs are described as being less neophobic than cats. This means that they experiment with novel foods—sometimes paying a heavy price in terms of gastric health—whereas cats avoid the unknown. But when it comes to water, dogs seem to enjoy the familiar more than the novel. As long as it is not heavily chlorinated, water straight from the tap is often favored over standing water. This may be an innate response that helps dogs avoid water that may be carrying persistent saliva-borne diseases such as distemper. Cold water may be preferred to warm water but chiefly when the dog is trying to cool off. So you are doing your dog a favor if you carry a vacuum flask of cool water for his refreshment during car journeys in warm weather.

Fun is the optimal reward

More than anything else dogs enjoy having fun—it can easily eclipse food as a reward, and many of us should use it in our coaching more often than we do. Unlike Uncle Wolf, an older dog still enjoys play. Running is fun, jumping onto beds is fun, jumping off beds is fun, and playing with other dogs is the best fun of all. Cajoling a playful response out of another dog is a particular joy. The way a young dog teases his elders speaks of an active if not sophisticated sense of humor. Indeed, the scientific study of dog laughter is now a very serious business.

The fun of racing ahead of the pack to get somewhere or to reach something first is difficult to overestimate. So I always feel a twinge of sympathy for little dogs in the company of big dogs, since they must resign themselves to never being the race winner. The small ones often redirect their enthusiasm for such pursuits towards an impressive attachment to their owners that confirms their value as a sanctuary when all around is teeth and legs and massiveness. The same can be said of older dogs that replace the thrill of the chase with the pleasure of the plod.

Dogs that lose all games of possession generally become less competitive. Those that get bullied during play become withdrawn. In general, opportunism and finding fun go hand-in-hand with optimism. There is emerging evidence that animals kept in frustrating environments behave as if they expect to find fewer rewards in life than those kept in appropriately enriching environments. This means that they are effectively less optimistic—the glass is half full—and less likely to offer new responses. For dogs, I sense that this can manifest as a lack of creativity when it comes to solving problems. For me, this is an argument against formal training sessions and a call for integrating training into everyday living. The more we turn our dogs' lives into one giant enriching opportunity, the easier they are to train.

Happiness is a tired puppy

Exercise out of the den is exciting because it offers so many opportunities: to "play, pee and poo," explore and eat food, meet and greet, mark and have sex—in other words, do most of the things that never arise as opportunities in the den. The excitement is greater than Feral Cheryl and her pack might experience at the onset of hunting forays because those dogs already have their liberty, and they can explore without boundaries. And to an extent the mission is possibly more exciting for domestic dogs because their free-ranging cousins rarely leave the home range. They have no cars that can take them to the "beach that is freshly washed" or the "forest that smells like no other."

The first step onto the grass of a park reminds owners how exciting a walk is for a dog. There is a great urgency for the dog to be off the leash and do all the things that have to be done: sniffing, marking, socializing, rolling and running. We get confirmation of their importance, but do we really know why these opportunities are so valued by dogs? We can imagine that rolling, especially after swimming, is somehow akin to rubbing oneself dry with a luxurious towel. But in a world of smells, it is an activity that may be critical in spreading odors. So, it's worth reminding ourselves that dogs find many activities intrinsically rewarding in ways that we will never understand. Consider chasing and pulling. When a husky pulls a sled or a border collie chases a sheep (or a jogger or a cyclist), there is no need for rewards. The activities are innately rewarding. For certain breeds, they are primary reinforcers. For the sled dog, the thrill of the activity is enhanced by novel surroundings and the ever-present possibility of encountering a sleepy rabbit. For the border collie, the worst outcome is that he is commanded to stop chasing the sheep, which is what shepherds do to punish unwanted responses such as biting the sheep.

We underestimate a dog's need for activity at our peril. As you can see in the table below, Professor Danny Mills, of Lincoln University in the UK, has published data from a survey of more than 500 dog owners showing a mismatch between what our dogs and we regard as appropriate activity levels.

Perception of appropriate activity levels

Reported behavior	% reported
Initiate interaction with the owners	81.4
Watch the owners all the time	47.5
Interfere with the owners when they were doing something unrelated to the dog	42.1
Very playful	72.8
Easily overexcited	48.2
Restless during the day	12.3
Restless during the night	8.9

For me, these important statistics point to a misunderstanding of what is normal. If more than 70 percent of dogs are very playful and more than 80 percent take an active role in initiating activity, those that do not are abnormal. These figures also suggest that many dogs are frustrated by lack of exercise, and pester their owners (with varying success) to make life more fun for them. Many of us struggle to meet our dog's appetite for play, and that's when we need to accept that the best toy for a dog is another dog. And that doesn't mean buy another one. Consider finding a suitable neighborhood dog and arranging for them to play together. As the immortal Snoopy once said, "Happiness is a tired puppy . . . "

Dogs are social animals

Animals that have evolved to live within groups seem to enjoy being stroked and groomed more than members of solitary species. Most socialized dogs, for example, enjoy being petted. This may reflect the social nature of *Canidae* (the scientific name for the animal family that includes dogs, wolves, foxes, jackals and coyotes) but, in contexts other than courtship and parenting, dogs groom other dogs (allogroom) less than cats or horses groom their fellows. Given the close proximity of canine group members in a pack, especially at rest, this is surprising.

Many dog owners acknowledge that it is a great joy to discuss private matters with their dogs. It seems unlikely that dogs gain much from these fireside monologues other than one-on-one attention. But they seek our company regardless of discomfort, following us into inclement weather, strange places and unpleasant crowds. They often move closer when we are distressed. So what is it about attention that is so important? Ask yourself what a dog can expect to gain from the company of another dog, and you'll come up with some interesting ways of providing the sort of companionship he has evolved to thrive on. Warmth, comfort, surveillance and membership of a team are important to consider here.

People who enjoy the companionship of warm-blooded animals, such as cats and dogs, frequently stroke or pet their animals. This may have benefits for both participants. During positive dog–human interactions, such as gentle scratching of the body and ears, concentrations of enjoyable, naturally occurring chemicals, such as endorphins, oxytocin, prolactin and phenylethylamines, increase in *both* species.

Of course there are a number of things that canine companions cannot provide one another, for example, chest scratching, smiles, tidbits, ball throwing, idle conversation and confidences. Although its value is debatable, we may use physical contact, such as petting, as a reward for desired behavior, as an alternative to offering food. This is regularly practiced by people who want to reward horses without using food. By scratching the horse at the base of its neck as another horse would, an interaction that is calming and rewarding, they are effectively hijacking a naturally rewarding response.

Grooming: transporting your dog straight to heaven

Anecdotal reports suggest that many companion dogs prefer to be groomed in certain areas, such as the front of the chest. Despite this, humans tend to stroke the top of the head and down the neck. It is believed that in some dogs, contact in these areas may result in displays of aggression because dog–dog contact with those regions (and especially the shoulders) has been related to attempts to assert status.

In one of my recent projects, I tried to determine the effect on a dog's heart rate of grooming in different anatomical areas. I was hoping to better understand the effect of human touch on our canine companions. I'd been discussing with colleagues the best way to make physical contact with a dog. We were mainly interested in this from a practical perspective, to work out how best to advise children and "non-doggy" folk on the topic. In our preliminary discussions we agreed that although certain under-groomed and often allergic dogs learn to position themselves so that they can be scratched towards the end of their backs, for most other dogs there seemed to be something generally magical about the chest area, the zone between the collar and the two front legs. Being tickled, scratched and groomed in this spot seems to transport most reasonably confident dogs directly to heaven.

Why might this be? Are they unable to attend to itches in this part of their body? Possibly, since even when completely curled round for some serious auto-grooming (scratching to the nontechnical) the hind legs struggle to reach the furthest parts of the so-called thoracic inlet (either side of the breast bone). Or maybe being touched here affects the flow of blood to the heart and simply makes dogs feel light-headed? The veins in this part of the dog are among the most exposed of any animal so this is possible, but why would dogs subject themselves to light-headedness when most of us don't enjoy it?

Some dogs hate having their faces touched (and almost all hate blowing—an important tip for children who might wonder why whistling near their dog's face makes it snap at them), while others get really irritable when their feet are handled. In contrast, the average dog just loves being petted, scratched and tickled in his hard-to-reach places. And yes, dogs can roll on their backs to relieve an irritation lying above the level their hind legs can reach. This point is critical. A dog depends heavily on its hind legs for grooming areas that are especially hard to reach. No wonder the worst flea-bite allergies manifest in one of the trickiest-to-reach places: just in front of the tail.

Of course, the extent to which dogs are rewarded by physical contact depends on their socialization with humans in general and their relationship with the people who are grooming them in particular. Indeed, acceleration of heart rate has been observed when someone who had previously punished a dog petted it. If the groomer and his technique are rewarding, then of course grooming and physical contact with the groomer are a resource. We are all familiar with the apparent jealousy that one dog shows when its owner makes a fuss of another dog. Attempts to intervene by slithering between the groomer and the "groomee" are far from subtle and sometimes spill over into aggression, quite possibly because there is so little room to maneuver that threats are poorly transmitted (among the tangle of heads and necks and hands and knees) and therefore go unheeded.

The duration of grooming and petting by humans is probably greater than any canine self-grooming bout. The effect of physical contact on heart rate may constitute a reward but only after some time. So, the immediate reinforcing effects of physical contact are likely to be secondary to other benefits, such as being generally close to social affiliates (friends). That said, grooming will always improve appearance and hygiene. This is particularly important for older dogs. Although the temptation may be to leave them alone as they snooze on the veranda, five minutes spent grooming away matted hair seems to pay enormous dividends in making them more presentable and comfortable and possibly even dignified.

Relaxed and comfortable

Do all breeds have the same need for sleep? Do dogs sleep when they are sleepy? Or do they use sleep to escape frustrating environments, for example, when the pack is away. This seems to be an acquired skill that some dogs take much longer to master than others.

Dogs relish comfort and seem to want others around them to know it. I am thinking of the ubiquitous sighing that comes from dogs of any age when (they reckon) they have found the best place to lie or the best way to lie in a place that can never be altogether comfortable. The sigh is generally emitted moments before the last opening of the eyes, as if to check that all is well before truly relaxing.

Conking out on the spot is a feature of puppy behavior, while turning around prior to lying down is characteristic of older dogs. The rotations seem haphazard and clumsy as they lower the dog to the ground, but they have emerged through evolution as a useful prelude to a good night's sleep—they are less common during the day, except in very senior dogs. The function of the spinning is the source of some debate, with some arguing that it is a means of checking for snakes before lying in their vicinity, while others insist that it flattens the grass in a single direction to form a rudimentary nest. Maybe it even smooths the hairs of the dog's coat and thus makes for more comfortable repose. Whatever the truth, the rotations generally end so that the dog's head is pointing vaguely uphill or in the direction of incoming traffic, such as at a doorway. Both of these outcomes would help the dog to rise rapidly to an emergency.

Other forms of comfort for dogs are easily overlooked. Try taking the collar off your dog when he is secure in the house and there is no need for him to be restrained or identified by any discs that may hang from his collar. You'll probably notice that he is quick to relax, and you will also spot how pleased he is to have the freshly exposed parts of his neck rubbed. The weight of a collar is negligible when first encountered by any animal but it's worth considering its accumulative effect over the days, weeks and months for which it is usually worn. In the same way, you might find removing a chain or necklace from your own neck disproportionately relieving.

Dogs love to work for a living

Once we have a clear idea of what dogs value, we can turn the topic on its head and ask: *What will dogs work for?* All of the valued items in this chapter can be used to pay (reward) dogs. If you hold the check book, you write the checks, and this makes you the leader. Which begs the question: *What do dogs regard as work?* The trite answer is: *Anything that gets them paid.* If nothing in life is free, everything is worth something.

Humans given dull, repetitive jobs resent work. But instead of being heavy labor or boring drudgery, the tasks we set dogs can be tremendous fun. Imagine a cushy, rewarding job that involves a wide variety of duties, such as sitting quietly at home, going for a jog with your boss, arriving at the water cooler, picking up toys, waiting briefly before playing with your buddies and, best of all, just relaxing. That is the sort of work well-trained dogs are asked to do. They have been trained to perform these tasks and just love the deal that has been struck by their owners. As opportunists, dogs relish discovering new ways to exploit their environment, social groups and leaders. Enlightened owners know what their dogs want and create opportunities for their dogs to acquire the most valued commodities through trained responses. This means that dogs are effectively working for a living. If rewards match their needs, most dogs will become workaholics.

CHOICE CUTS

- The harder a dog works for certain resources, the more he values them.
- Food and drink, fun, company and comfort seem to be key resources for dogs of all ages.
- Like many social species, dogs value being groomed.
- A bowlful of food is a bowlful of training opportunities.
- For some resources, dogs may be prepared to pay any price.
- More than anything else, dogs enjoy having fun.
- Dogs have evolved to compete with their group for all the good things in life: Excellent coaches tap into these needs.
- The best toy for a dog is another dog.
- Happiness is a tired puppy.
- Dogs love to work for a living.

Play-fighting among dogs depends on bite inhibition. Tinker and her son, Neville, have developed highly sophisticated play styles that feature relentless mouthing.

Tug-of-war games with adult dogs help to build a pup's confidence. In this case, Ben plays with his daughter, Nessie. It is interesting to speculate whether related dogs have similar odors that help them recognize kin.

4

What Dogs Dislike

In this chapter, we'll look at how a dog can have a bad day.

Bad eating habits

While most dogs have a passion for food, for some it can also be their undoing. A dog with an enormous appetite can make disastrous mistakes when choosing what to swallow.

DOGS WILL TRY TO EAT ALMOST ANYTHING

Veterinary surgery professors around the world tell amazing stories about what they've removed from the esophagus, stomach or intestines of dogs. The old favorites are balls, pebbles and sticks, ill-judged corks, corn cobs and tasty barbecue skewers. But there have also been the downright dumbfounding—from corsets to spectacles and mobile phones (even when they were the size of house bricks).

The objects dogs consume defy our reason, since to the human eye they don't at all resemble a prime rump or an oven-roasted chicken. The fact is, many of these items carry rich quantities of human scent, and dogs may be selecting these over less smelly but more digestible items. Perhaps it is their owner's odor on the item that proves irresistible to the dog, or maybe it is their novelty value or that they seem so important to the social group's leader. Either way, the items rarely function properly after being retrieved from gastric liquor or fetid bowel contents and can certainly threaten the lives of their consumers.

Professor Danny Mills' survey (see table on page 51) of more than 500 UK dog owners showed a prevalence of apparently unorthodox dog appetites. Some of the items swallowed can be an ongoing problem. For example, all companion animal vets will confirm that stones are surgically removed from the same dogs year in and year out. Each operation brings the risk of infection, and each additional breach of the adhesions increases the risk of more adhesions. It is worth noting that since Professor Mills based his study on voluntary reports from owners, his data may well be underestimating the true prevalence of the tendency for dogs to swallow almost anything. Certainly, if they are observed for long enough, most normal dogs will be seen eating grass.

DOGS TEND TO OVERINDULGE

If eating the wrong things can be disastrous, so can eating too much of the right thing. The stomach volume of a dog represents 60–70 percent of the total capacity of the digestive system. Dogs eat first and ask questions later. They have evolved to gorge themselves and then snooze as the job of digesting the food begins. Wild dogs often hunt only once every five days because rapid consumption of a carcass means that they have reserves on board and, equally, that they are too full of food to hunt successfully.

Unorthodox items consumed by companion dogs

Item eaten	% of dogs reported
Another animal's feces	29.2
Unusual foodstuffs	11.4
Stones	11.1
Own feces	5.3
Another dog's feces	4.4
Paper and wood items	4.0
Grass and soil	3.5
Plastic and glass	2.6
Clothing and furnishing	2.4
Coal and charcoal	0.6
Insects	0.3

Dogs rarely vomit when they have eaten too much. They have few mechanisms to counter overindulgence, which creates potential hazards, such as bloat. This is a deadly condition, especially for breeds with deep chests, which are likely to accumulate gas from fermented food that can twist the stomach. This locks off the stomach's blood supply, causing extreme pain and damage to the stomach. Stomach bloating and twisting (gastric dilatation and torsion) is a critical veterinary emergency.

We see here how anatomy and physiology influence behavior, which is important if we want to use food as a reward. As they ease themselves away from an empty bowl, most dogs just want to kick back and relax—they certainly won't be motivated to improve their heelwork or fine-tune their agility through the weaving poles—so to offer them food-based rewards after a meal is pointless. You are asking them to work for the right thing at the wrong time. The best time to train your dog with food is shortly before feeding its regular meal, when its stomach is empty and its digestive juices are flowing.

It's then primed to work for food, just as Uncle Wolf was when he gave his best in the chase to bring down his dinner.

DOGS DON'T CRAVE VARIETY IN THEIR FOOD

As mentioned earlier, variety in food is not essential for dogs. Food spends so little time in a dog's mouth before being swallowed that it's hard to bore a dog with a certain type of food. Having said that, the opportunistic nature of dogs makes them keen to try new foods.

CHEW ON THIS

Boredom is very much a human construct. It's based on the notion of an unfulfilled mind and therefore depends on evidence that animals have minds in the first place. Sadly, this may seem obvious but is far from easy since it raises the notion of consciousness. Is a bored animal aware that it is under-stimulated? It is very difficult to measure this change in an animal, and maybe it's not even important. Maybe all we need to know is that being under-stimulated is in itself frustrating. So a dog that is unable to dig in mud or play with other dogs or exercise its limbs may be described as frustrated even without brain-wave studies to show that its brain isn't being stimulated.

FOOD IS NOT THE ONLY REWARD

A satiated dog can be a disinterested dog, and dogs trained solely with food can be challenging for two reasons. First, they can become difficult to "jackpot" because they expect food (and tasty food) at all times. Second, and more commonly, they can be completely unfocused when they have just been fed. The message is clear. We should use a variety of rewards to keep dogs interested. Once we know whatever nonfood resource motivates them most, we can focus on giving that reward for the responses we are training. We can also align training with the target behavior. For example, take a dog that loves a chase. To train him to dig, for example for truffles, we might use a ball laced with truffle oil and hide it under some fallen leaves. The target behavior is displacing the litter to uncover the article. Once the ball is discovered, we throw it and thus provide a chase game.

Malnutrition

We'll look at obesity in chapter 6 (Sex, Disease and Aging), so for now let's concentrate on what else we may be doing wrong when energy input is out of balance with output and when our dog's diet is lacking in critical elements.

HUNGRY DOGS

Despite the number of obese dogs we see all around us, many dogs stare at food being scraped into the kitchen trash bin and seem to be asking what the bin has done to deserve such rich rewards. Of course, in developing countries there is often insufficient food for the human population, let alone their animals. In such marginal situations, pets are a luxury very few can afford. On the other hand, in developed countries, hungry dogs may be the result of overzealous dieting imposed by owners seeking to treat or prevent obesity. This is especially the case in spayed bitches because they use food with great efficiency. Predisposed to obesity, such dogs have a formidable appetite that impels some of them to raid garbage bins and even escape from their homes in search of food.

Owners also sometimes restrict their dog's food to maintain motivation for food used in training. This is not necessarily tough love because as we will see in chapter 15 (Working Alliances) it is commonly used without any adverse effects in training sniffer dogs.

Feeding dogs together in a group can make it very difficult to monitor each dog's intake. Competition within the group may leave some dogs hungry, since they tend to hoard and defend any foods, such as bones, that can be carried. So, if you are feeding bones, aim to put out more than can be monopolized by one individual dog.

FEEDING THE WRONG FOOD

Ignorance and miserliness may cause some owners to feed the wrong form of food to their animals just because it is cheaper. With examples from my own clinical practice that included an instance of cat food being fed to a pony, it is easy to see why the consequences can be disastrous.

Overfeeding young dogs with calcium can precipitate growth plate disorders. Meanwhile, giving all that lard from the roasting tray to your dog after Sunday lunch can cause pancreatitis, an acute condition that commonly presents with the dog screaming in pain. We do well to think twice before assuming that what we eat will suit our dogs. Misguided feeding of cooked onions and grapes to dogs can cause hemolytic anemia and renal failure, respectively.

Orphaned newborn animals of any species are likely to become malnourished because the balance of fats and the types of sugars in the mother's milk of any species is very particular. So, it is wrong to assume that all young animals will thrive on cow's milk. Giving the wrong formulation to newborn pups can cause life-threatening diarrhea and extreme dehydration. Commercial milk replacers are commonly available from veterinary clinics but tend to be used only after emergency measures have been administered in the middle of the night. These emergency measures sometimes create an emergency in their own right. Anything from cow's milk to brandy can be ill advisedly dribbled into the mouths of newborn pups when their owners are reluctant to seek professional help.

DOGS NEED PLENTY OF DRINKING WATER

All codes of practice for the care of animals stress that clean water must be available at all times. That said, the gruesome turbidity of the water dogs drink from puddles surprises most of us. Even though the water must be teeming with horrors, dogs rarely fall ill as a result of drinking it. This serves as some sort of testament to the strength of dog stomach acid and its ability to deal with bacteria.

Although on the face of it, it may seem fairly easy for dog owners to supply fresh water, it's worth exploring why problems still occur. Water may be provided but in contaminated bowls. Simply topping these up with fresh water does nothing to reduce their tendency to ferment an algal or bacterial soup. Water in clean containers may be knocked over by dogs or their tethers. It may be contaminated by misadventure (in the case of chained dogs, simply by the chain being dragged through their own feces and then the water). Dogs may urinate beside water and inadvertently urinate into their water bowl. Occasionally a dog can get especially thirsty as a result of acute disease such as diarrhea, or increases in the salt content or decreases in the water content of their food. On the other hand, you may find that your dog is urinating in the house overnight; you might remedy this by taking away its water bowl in the evening.

When a dog doesn't have enough drinking water, blood thickness (plasma osmolarity) is increased and blood volume is reduced (hypovolemia), both of which are signs of dehydration. The animal will initially feel thirsty, but if it fails to drink may then become weak as its body starts to malfunction. The kidneys are especially susceptible to such problems. Fear of kidney disease is the main reason I advise against ever restricting access to water.

Resources under threat

When a dog's resources are gathered together in a household, they're worth defending. It's this motivation to defend resources that seems to cause the most aggression toward humans in the home. The table below, with data from more than 500 owners sampled by Professor Danny Mills, shows the percentage of dogs that were reported to growl, snap and bite at humans.

The distribution of targets of aggression by companion dogs

Target	% of dogs
Unknown adult male visitors	32.3
Unknown adult female visitors	26.4
Unknown male child visitors	14.7
Unknown female child visitors	14
Familiar adult male visitors	6.6
Adult male household members	6.2
Familiar adult female visitors	5.7
Adult female household members	4.3
Familiar female child visitors	4.2
Familiar male child visitors	3.5
Male children in the household	3
Female children in the household	2.5

This table helps to explain how certain types of humans create more of a threat than others. Familiarity, gender and age all have a bearing on the perceived threat, and this isn't necessarily because dogs have learned that these are predictors of meanness. The table suggests that something innate makes dogs expect incoming unknown adult males to create trouble and threaten resources. Of course, socialization is the best way of breaking down this prejudice. Puppy owners should be encouraged to socialize their dogs with plenty of unfamiliar adult males. Perhaps we should be paying grumpy old men serious money to sit in the corner of puppy classes and dole out pigs' ears, dried liver treats and sliced sausage!

Of course, although many of us value dogs for their ability to alert us to incoming visitors, threats do not necessarily come from outside the den. Members of the social group who, after all, are especially aware of the available resources can threaten injury or displacement to acquire them. A dog's closest allies can also be competitors, so a great deal of time and effort in dogdom is spent in play and appeasement behavior to reduce the expression of threats such as these. It's a shame that because we can't detect the nuances of canine body language, we miss out on a great deal of our dogs' attempts to keep us sweet. It is also likely that when we see trainability and regard it as a willingness to please, we are significantly underestimating our dogs' social skills.

Experiencing pain

Dog owners are understandably concerned when their dog is in pain or discomfort. However, the difference between pain and discomfort is poorly defined. Veterinarians may describe the same dog as undergoing pain one day and discomfort the next. Sometimes this is a response to an owner's need to hear the comforting thing. For example, an owner concerned that her dog needs pain relief may be advised by the vet that the dog simply has "discomfort." When helping an owner make a decision to euthanize a dog, vets often use the word "pain."

A dog has pain receptors (nociceptors) all over its body. These help to keep the animal away from extreme heat, cold, and other irritants. They work in an instant via reflexes and over the longer term via learning. And, of course, pain is helpful. It teaches pups what to avoid in life. By way of a human example, those born without nociceptors usually don't live very long. If fractures and burns don't hurt, the results can be devastating, and if the larynx doesn't have the receptors to notice that something is blocking our airway, we are sure to choke or drown.

Modifying behavior

As we will see in chapter 11 (The Artful Dodgers) negative punishment, or omission, can be used to improve or modify a dog's responses. Animals being trained in some new behavior will first attempt to use an established response. If not rewarded (reinforced) at that point, they'll be less likely to repeat this now-unwanted response. Reinforcement has been withheld so the animal has been negatively punished. This makes the dog more likely to respond in new ways. The trial-and-error process continues.

ELIMINATING UNWELCOME BEHAVIOR

The use of training discs, as developed by the late British behaviorist John Fisher, relies on omission, or what can be regarded as a secondary punisher. These training discs make a sound when rattled or thrown to the ground. They are introduced to the dog together with the removal of food that the dog is expecting to get. This is done three or four times, so that the discs become strongly associated with frustration. Once the link is established, the discs can be thrown (*near the dog, not at it!*) when unwelcome behavior, such as barking in the car, occurs. Introducing the discs can stop the response for a brief period and give the trainer an opportunity to reward the dog for stopping. In a similar and much more general way we can use the command "No!" as a secondary punisher that is effective because it's used together with omission of something positive—such as praise or attention—that the animal would normally expect.

For some trainers and veterinary behavior specialists, the concept of punishment is too closely married to one of physical abuse for them to admit they use it. Unfortunately, they are allowing political correctness to muddy their thinking.

WHEN A DOG LOSES A LEARNED RESPONSE

Most of us associate *extinction* with dinosaurs and dodos, whereas for learning theorists, the term refers to the disappearance of a learned response. Surrounded by dribbling dogs, the Russian physiologist Ivan Pavlov found that if the sound of a metronome was paired with food, it would continue to make the dog salivate, just as long as the distinctive sound continued to be followed by food. If the metronome was sounded again and again but it wasn't followed by food, causing the dogs to stop salivating, then the process of *extinction* occurred.

Extinction applies to all the examples of classical conditioning. So, if clicker trainers do not reliably link the sound of their clicker with food, the clicker stops working. By the same token, if rewards stop appearing, animals will no longer offer trained behaviors to get them.

Fear

Once an animal has learned to fear something, it will show fear of similar things. If a neutral stimulus is paired with a shock and then presented repeatedly without any shock, it will lose its ability to evoke fear. Dog owners who use invisible electric fences to contain their dogs can switch the power off occasionally, but they cannot dispense with it completely, or the dogs will lose their fear of the barrier.

Fear of the potential consequences can be fundamental to a dog's behavior. If dogs have learned to associate pain, discomfort or disappointment with certain stimuli, they may well learn to avoid these stimuli, or, if they can't avoid them, to fear them.

Dogs may fear loss of resources and possibly even loss of rank. That said, for a dog to be aware of its rank, it probably has to be aware of others' rank, since rank is a relative term. The concept of benign leadership is critical here, since most experienced dog folk will confirm that they have known pack leaders who never have to lift a lip in anger. This serene, passive model is what human life-coaches of dogs should aspire to.

Rank in itself may be a nebulous concept for dogs, but what matters to the likes of Feral Cheryl and, to a lesser extent, domestic pets is how rank affects their access to resources. If a dog loses its access to more and more key resources, it is likely to become rather unhappy. For each dog, the value of each resource varies from one context to another and from one time to another. So the motivation to get and keep certain resources is never constant. It's wrong to assume that a lead animal will always insist on having access to all the resources all the time. Instead, we should understand that high-ranking dogs are those that have access to most of the resources most of the time.

Humans can get bitten when they ignore warning signs and reach for a resource that a dog is determined to defend. Obviously, we can't ask our dogs how possessive they happen to be feeling towards a particular resource, so it's safer to assume that all critical resources will be stringently defended. From this point we should strive to be able to take all resources from the dog without provoking a defensive reaction. For example, as a measure to prevent my dogs from biting visiting children when they are guarding a bone, I train my dogs to surrender their bones to me only briefly and always for a jackpot reward. But perhaps the best strategy when children are visiting is to remove all bones that dogs may become possessive over.

Deception

If the concept of rank is tricky to study, then measuring a dog's ability to deceive is even trickier. An example might be dogs that creep onto their owner's bed at night but do so only when the owner is asleep. They seem to deceive because they give the impression they are content to sleep elsewhere but stealthily slither onto the bed when not being watched. So do they have some understanding of the surveillance that an owner needs to detect their approach? Or do they look and listen for the breathing patterns associated with real as opposed to feigned sleep and use these as a cue to jump on the bed without fear of detection?

Another example is dogs that bark at the door when there is actually no one there. By raising the alarm they may effectively lure other dogs in the household away from key resources and race back to monopolize them. The all-important questions here are whether they know what the other dogs are thinking or whether they have simply learned that barking at the door has desirable consequences in another part of the house.

It's important to realize that distractions have a finite shelf life. They work only for so long because the victims of such deception become canny to falsehoods. It's worth understanding this because the latest advice for owners as they deal with unwelcome behaviors is to distract the dog. This makes some sense, since it's hard to reward other behaviors (counterconditioning) if the dog finds unbridled pleasure in its new, inappropriate activity. The problem is, the more we try to deceive and distract our dog, the less willing it will be to trust us.

Here is an example. Owners faced with aggression from their own dogs over access to sofas or beds are often advised to go into another room and create a distraction: jump up and down, play with an imaginary toy, squeal and giggle. These novel activities may distract the dog once or twice, but very soon they lose their novelty and simply tell the dog that he has successfully got the sofa or bed to himself. Dogs seem to work out pretty quickly that their owner's antics are intended to displace them rather than simply entertain.

Punishment

The debate around the use of punishment in dog training has been derailed by a misunderstanding of the term. Punishment is *not* physical abuse. It is anything that reduces the likelihood of certain behavior in the future. Strictly speaking, saying "No!" qualifies as punishment if it discourages certain actions. The non-punishment movement in the dog-training world emerged out of a justified desire to prevent hands-on abuse of dogs. However, now it finds itself saying "Never say No!", and I find myself asking, why throw away half your toolkit? For my dogs, "No!" has always meant "Stop doing that because it will not help you get a reward." This makes the unwelcome behavior less likely in future; therefore it is punishment but not abuse. Just like a red traffic light, it helps the dog identify behavior that isn't worth pursuing. All of this is important because it shows how the inexact use of terminology can confuse an issue.

Let's look at how this works with fear. The politically correct approach is to say that dogs are never aggressive, they are just frightened. They are never angry, they are just fearful. They are never offensive, they are only defensive. I can see merit in this view since it may stop people from hitting dogs over displays of fear-based aggression, but has it helped us understand dogs and their training? Taken to the extreme, fear can excuse almost any behavior if one insists dogs are only ever defensive and never aggressive. For example, the dog that growls when its owner tries to get it off the bed is really just defending its position on a comfortable sleeping place; a dog in a yard that bites a passer-by may be excused on the grounds that it was protecting its territory; a dog that nips a child while chewing a toy was simply fearful of losing its surrogate bone. So fear of loss can be used to justify potentially dangerous responses. However, there are problems with this approach, since it blurs the distinction between a dog's expression of fear and an assertion of possession or status. Once dogs have used their teeth to defend a resource, they are unlikely to stop. They either learn that humans are scared of them or are scary; either way, aggression becomes more likely. Clearly, a central strategy here is to avoid clashes and flashpoints that pit dogs against humans. We don't want dogs learning that they can displace us since being able to do so can cross from one context to another and, in the process, can align with the characteristics of high social rank. The problem, of course, is that very few owners appreciate the definition of punishment, much less the unwanted baggage it brings with it. When distractions or mild aversive stimuli fail to make the behavior more likely in the future (i.e., to punish it), novices fall into the trap of escalating the aversiveness of the stimuli and become aggressive themselves. That is why it is generally safer to assume that dogs are fearful, not competitive, and that aggression should be extinguished only with positive reinforcement. That said, until we accept the importance of social order in a dog's world and study its implication in the shared domestic context, we will be missing the critical details that make some dog–human pairs work vastly better than others.

Shame

Many owners describe the look on their dog's face when it is caught sleeping in a forbidden area, raiding the garbage or trashing the "wrong" leather item as one of shame or guilt. It is tempting to assume that these emotions are direct equivalents of human value systems, but this may be selling dogs short, as well as making too many leaps of faith. The last thing dogs would want to do is annoy us for the sake of it. What use would this be to feral dogs or wolves? It seems entirely unproductive, and, at the planning stage, it depends on the dogs knowing how we feel about the transgression they are about to perpetrate. So, for the peccadilloes to have maximum effect, the dogs must appreciate that the sleeping area is a duvet that has just been washed, that the raided garbage represents a mess and a chore for the owner to clean up, and that the item made from dead cow is Italian, expensive and designed for wearing on one's feet.

Leading scientist Vilmos Csanyi from Hungary has studied emotions in dogs. He argues that shame is not a canine emotion since it relies on very complex social relations, relations that dogs have never needed to get along with humans. I would argue that jealousy, cheekiness and mischief are in the same bag. They appeal to owners who want to explain every behavioral outcome, but ethologists remind us that these are not behaviors but abstract ideas or constructs.

CHOICE CUTS

- Dogs adopt a trial-and-error approach to food selection.

- Eating the wrong food can get dogs into serious trouble.

- Threats, pain, discomfort, and the disappointment of non-reward are all bad news to dogs.

- Using a variety of rewards keeps dogs interested and well-trained.

- Ensuring a regular supply of drinking water will prevent dehydration and disease.

- Dogs sometimes become aggressive when they perceive a threat to their resources.

- Being consistent with rewards will ensure that your dog maintains trained behaviors.

- Punishment is *not* abuse. Simple vocal cues, calmly delivered by owners, help dogs identify behavior that is unproductive.

- There is little evidence that dogs feel what we call shame.

- The tasks we set dogs can and should be tremendous fun for them.

Often labeled a displacement behavior or a calming signal, yawning is sometimes a sign that dogs are slightly uncomfortable. Being inherently cautious, Tinker often yawns when she is the focus of human attention; e.g., when having her portrait taken.

It is tempting to say we can see guilt on Annie's face as it pops out of a hole she has just dug in the garden. In reality, it is more likely that we are simply seeing appeasement signals that we have unwittingly trained.

5
Networking Among Dogs

In this chapter, we'll look at how dogs relate to one another and then climb into the debate over how this may help or mislead us when applying dog–dog models to dog–human and human–dog relationships. Generally, so far, we have been clear that where there is no dispute over resources and dogs simply defer to us, this deference may be as subtle as getting out of our way, something we scarcely notice and hardly ever reward. The fact that we receive deference and do not demand it marks us out as clear leaders. In addition, we hold the resources and, using them, can train all the behaviors we regard as desirable. This is something other dogs cannot do as effectively as we do. Our relative height means that dogs automatically look up to us, literally and figuratively. It is possible that all of this makes us super-dogs: initiators of expeditions, grooming, play and feeding and leaders who would never be worth questioning. Dogs' motivation for group membership and preference for the peace that comes with a clear leader is clearly matched by the ease with which we can deliver. So we need never stoop to conquer. What a privileged position we find ourselves in! How remarkably easy is it for us to exploit this captive audience! How sad it is that this gift is ever abused!

We do well to study the ways in which social order is created and maintained in stable groups of dogs. The peacefulness that usually defines such established dog communities reminds us that there are very few breaches of social order and that aggression is rare. This is underpinned by clear signaling and deference, delivered not demanded. Rough physical contact is far more often part of play than of violence, and it is foreshadowed by strong signals.

Social groups in feral dog contexts are much more stable than they are in the human–dog domain. Feral dogs do not meet strangers on a regular basis, visit parks or go on holidays. Dogs have not evolved to know that the new social groups that arise in the company of strangers, on a visit to a park or a trip to a holiday destination, are not going to last for the rest of time. Making sense of these novel groupings either relies on learning gradually about each resource and who is allowed it and when, or it involves some social ranking (status) that removes the need for constant disputes. There seems to be merit in working out swiftly and painlessly who must defer to whom. Scientists generally base their measurements of social hierarchies on who can displace whom from food and, less often, on who can initiate contact with whom. The question is whether such social order in dogs can include humans and whether it may explain how humans occasionally get bitten.

Whether dogs have evolved to use their amazing tools of social order with another species (most notably us humans) is contentious. Some say the so-called pack theory that emerged from a limited number of studies of managed wolf communities provides the template for dog social life. In contrast, others argue that the dog's ability to socialize with humans, a species completely alien to them, makes them entirely different from wolves. The middle ground suggests that dogs are likely to rely on the social repertoire they have evolved to use with other dogs unless they have been socialized with other species. Certainly dogs that show the greatest behavioral flexibility, an attribute often labeled compliance, tolerance, and even forgiveness, are those that have been the most thoroughly socialized. The socialization period (in which puppies learn the stimuli that they can accept as familiar) permits us to see the most striking differences between dogs and wolves. Thorough socialization during puppyhood helps them to accommodate the cascade of novelty that defines the life of the modern dog. It may also help dogs learn that nearby humans are not using stand-over displays when they stand up and that other human behaviors, such as smiling and stroking, are in no way related to the teeth baring and shoulder presses issued by other dogs.

When dogs and humans come together under one roof, there may be many ways in which we inadvertently send inconsistent and mixed messages. We often fail to recognize how dogs value some of the resources humans happily share. For example, dogs may place more importance on a snug spot on the sofa than we ever will. We may inadvertently reward the small beginnings of resource-guarding and so may set dogs up for elevated aggression in the future. We may miss some signs of deference and fail to respond appropriately, ignore signs that tell us dogs are motivated to keep their distance, and touch dogs in ways that, without qualifying play signals, are downright rude. We stare into dogs' eyes, and stand over, reach for and cuddle them around the neck in ways that dogs use when threatening one another. Small wonder then that we push some dogs to defend themselves and their resources. Do clumsy interactions such as these also prompt dogs to defend their status?

As opportunity-giving species, humans are well worth being around, and most dogs succeed by exploiting us. If we appeared less like super-dogs and more like regular dogs we would soon see ourselves in the thick of the social strata. Imagine if we spent most of the time on all fours and ate off the ground. This would surely lead to more dogs defending access to resources and, yes, quite possibly lead to them displacing us more often than we currently displace them. But the fact is that, apart from the occasional human toddler crawling among the dog bowls, we generally don't make matters so blurred for our dogs.

Relationships between dogs

To get a sense of how dogs get along with one another, let's begin by considering the way pups relate to their family group. With their eyes sealed shut, pups are functionally blind for the first 10–12 days

CHEW ON THIS

In dogs, tissues around the nipples, and especially in the cleft between the left and right chains of mammary glands, secrete appeasing pheromones. Recent studies have confirmed that these pheromones play a critical role in helping pups locate the nipple, and effects in adult dogs have also been found. In both pups and adult dogs, these pheromones have a calming effect and may even be involved in the calmness that a pup experiences when it locates the nipple even before milk flows. Researchers have found that in adult dogs, a synthetic form of the same appeasing pheromone reduces signs of anxiety in vet clinic waiting rooms. The recent success of this product in calming fearful dogs, such as those with separation-related distress or thunder phobia, is extremely encouraging. Dog Appeasing Pheromone, as it is marketed, is also being trialed as a means of calming pups once they have been placed in a new home.

of life. Given that dogs live in a world of odors, we can assume that for a puppy the smell of its mother is extremely important. The odor of a pup and his littermates mix with his mother's scent to create a world of safety, comfort, food and warmth. Until his eyes open, a gradient of odors helps him find his mother and siblings. It is likely that the earliest lessons a pup learns about these odors stay with him for life. It's possible they even help him to recognize his kin after a period of separation.

Leaving home

Moving to a new home is very challenging for young pups. Dogs have evolved to either live within their family group or to disperse. Although many scientists now think of dogs and humans as having evolved together, we can't argue that dogs have evolved to be sent to live with a different species at the age of 6–8 weeks, any more than we can suggest that they've evolved to be neutered. For some reason, pups are especially flexible at this age and are less likely to clash with aliens in later life if they are introduced to their new social grouping at this time. Fundamentally, dogs are behaviorally flexible enough to live with another species. Some scientists call this preadaptation, meaning that the dog's biology favors domestication. The fact that some domestic dogs can turn feral within a generation strongly suggests that our selection pressure for domesticated traits has done little to compromise their ability to succeed without us.

Staying with the pack

If we want to look at the pros and cons of staying with the pack in the wild or separating, we need to consider how Feral Cheryl's brood fits within its family group; in other words, what's in it for the pack to stay together? True, fresh pups eventually bring all the benefits of any group members, including added safety, by increased surveillance, increased dog-power when fighting enemies, and more breeding opportunities. But at what price? They demand to be suckled and fed by regurgitation, something that, on top of the pain of chewed nipples, eventually creates sufficient conflict to force the bitch to wean her litter. After weaning, they may cooperate in many ways, but they also compete for food and, indeed, for all resources.

If they do stay around in the long term, the bond that develops between parents and siblings is very sophisticated. The orientation of the head during rest can bestow favor or deference, and the stiffening of a posture can send a serious warning. A glance from a high-ranking adult can stop an active youngster dead in its tracks, while a stare is very intimidating indeed! For the youngster, looking away can equate to bowing and scraping, and rolling over can signal the absence of any challenge.

Sibling rivalry

Sibling rivalry can be a serious issue among dogs. That's why, despite the occasional success story of two [dog] siblings spending their entire lives together, rearing pups from the same litter is not widely recommended. Puppies understand "puppyese" much better than English, and so they always relate better to each other than they do to humans. They will spend more time with each other than they do with any human caregiver. The human is in danger of becoming a food resource and nothing more. If the pups are of opposite sexes, they may thrive in this semi-feral fashion for the rest of their lives. Further, if they are the same sex, they become each other's closest competitors, play-fighting up to a point and then often fighting in earnest. They may compete for food, toys, sleeping places and the favor of their senior pack members.

CHEW ON THIS

Among free-ranging dogs, successful life-long pairings of sib-lings is far from common. Indeed, the rarity of such same-litter-same-sex alliances in nature should tell us something. Littermates struggle with one another from the word go as they compete to access the preferred nipples. But both milk-sharing and power-sharing are cumbersome and cannot be achieved with absolute fairness; one member of the alliance usually benefits more than the other. Brothers may adventure successfully together, warn each other of and defend one another against novel threats, and may even take over packs, but in the long run they never equally share the distribution of such valuable resources as mating rights.

In the domestic context, dogs rarely get to choose a sexual partner since this is the job of the breeder, and, sadly, we know that pursuit of certain traits prompts some breeders to mate brothers with sisters, father with daughters and mothers with sons. In the wild, incest tends to reduce the genetic diversity of a dog's offspring, which in the long run means that those dogs who are predisposed to mating with siblings will not to do so well in evolutionary terms. Dispersal in search of a mate allows new genes to exploit niches and possibly even to parasitize off another gene's previous success. For example, by moving into an established pack, a new breeding male has the advantage of the cultural successes, such as hunting strategies, that his new group has worked hard to establish over recent years.

Canine communication

The intricacy of canine communication—with deference by subordinates and chilling glares from superiors—means that it's hardly ever necessary to use teeth. And this is just as it should be, since to enter combat is to acknowledge that other strategies, including deference, have not worked and that the protagonists are of sufficiently similar resource-holding potential. Combat disturbs the fine balance within the group and increases the risk of injuries. However, although aggressive encounters as a result of competition are rare, contact among group members certainly occurs when there are few other distractions (such as prized possessions). Mutual grooming is a great example. In domestic situations it's more common in dogs that have grown up together than those that have been introduced to one another as youngsters or adults. As we've learned, preferred grooming sites include the back of the neck, around the ears and along the midline of the back. These preferences suggest that scratching of such hard-to-reach places is particularly gratifying, presumably as it releases oxytocin, the "hormone of love." It seems likely that by shuffling into a particular position, the "groomee" signals the groomer when he/she has hit the spot, and also when to stop.

Face-licking

Face-licking, focused around the mouth, ears and sometimes the eyes, is a dog's way of saying that it wants something. If a youngster licks its parent's mouth, it could mean it wants food. If it licks the mouth of any older dog, it might be acknowledgment, interaction or affirmation—offering submission after this sort of solicitation helps maintain the social order. If a male licks a bitch's ears, he seeks permission to make courtship advances.

Barking

Dogs bark more than wolves, not least because their environments stimulate it. Of tremendous importance here is social interaction, which explains how one dog barking can prompt others within earshot to do the same. We all recognize the way in which volleys of barking can travel across a neighborhood. Although dogs bark more than wolves, not all breeds bark equally. Basenjis, an ancient African breed, the so-called barkless dogs of Africa, are among the least

CHEW ON THIS

A recent study on barking has shown that the specific sound of a bark varies according to the context. The study identified three contexts that evoke different barking: play, isolation and disturbances. Generally, barks in response to disturbance last longer and are pitched lower than those used in isolation and play situations. Play barks are especially high-pitched and unevenly spaced. These findings are consistent with the work of Hungarian researchers who showed differences among the vocalizations of dogs that were alone, playful or confronted by strangers. This research group went on to show that a dog's bark can convey its emotional content to humans. Even when they were more familiar with dogs of a different breed, humans were reasonably accurate in identifying the context of these different barks. They correctly assigned despair and fearfulness to the alone barks, happiness to the play barks and aggressiveness to the warning barks from dogs encountering strangers. Perhaps breed differences in bark quality are a bit like human accents.

vocal breeds, whereas poodles and Shelties tend to be the most vocal. All barks are not equal though. Barks have different meanings, and humans are reasonably good at identifying the emotional state of dogs after hearing them bark. This may not come as news to those of us who've noticed dogs using different barks in different contexts.

Marking

Because they have evolved to occupy a defined area, dogs routinely patrol the tracks and significant landmarks within their domain and anoint them with their scent. They are especially moved to mark when they detect another dog's scent on top of their own. Using their urine or feces, it's possible that females mark around their nest to warn off intruders. Dogs will also mark leftover food and areas where food has been found.

Establishing contact

Most social animals touch each other regularly in companionship and to resolve minor disputes (pushing and shoving). Physical contact when you first meet a dog can be absolutely critical. The latest studies (specifically concerning dogs in shelters) show that physical contact, in the form of grooming, can establish a bond between a needy dog and a human stranger more effectively than training or free food. However, this finding doesn't justify forcing yourself on a dog that is clearly telling you to keep your distance. This research is significant because it shows that grooming not only makes a dog in a shelter more amenable to strangers (potential new owners), but the dog is also less likely to get agitated when separating from new humans.

CHEW ON THIS

My colleagues and I have studied the environmental and sex-related factors associated with urination and defecation. In our survey, owners reported that dogs "went" more often when they were visiting unfamiliar areas. We classified these areas as regular, occasional and novel, according to how often they were visited. Areas visited only occasionally were associated with more frequent urination and defecation than those that were visited regularly. But novel areas were associated with the highest rates of all. This suggests that the motivation to mark with both urine and feces drops in familiar areas. Our study also showed that the hormone profile of dogs also affected marking. Entire males were reported to urinate and defecate more than females in all areas and to defecate more than neutered males in regular and occasional areas.

Having asked the owners all these personal questions about their dogs' toilet habits, we went out and observed what was going on in parks. A colleague of mine volunteered to observe dogs as they were exercised off leash. She recorded the number of times they urinated or defecated and collected all the poo that they left behind. This study confirmed that entire male dogs urinated and defecated more frequently than other dogs.

When the feces of all dogs were analyzed, we also discovered that as the number of poos increases, the runnier they become. This is important because we established a link between the urge to mark and the risk of runny poo, thereby identifying a cause of diarrhea. Taken together, the two elements of this study suggest that fecal marking may prompt diarrhea in unfamiliar or particularly stimulating areas. The results also show that entire male dogs are at particular risk of this. Owners and veterinarians should take note and perhaps curtail this particular form of exercise when dogs have diarrhea.

Grumpy old dogs

Puppies usually get along famously, but older dogs are often less tolerant of others. Clear communication among established groups of dogs can give the impression of harmony, but disputes can simmer for years and boil over as a dog ages. For example, as a senior bitch approaches her dotage, she may become less subtle in her signaling and more emphatic. As she increases the power of her stares, glowers and lip curls, the time will come when she might decide to use her teeth to secure her resources. This may prompt antagonism, especially with other females, as the boundaries are tested. At this point her increasing physical frailty may be so exposed that her position becomes untenable. The brutality of fights between a senior bitch and her challengers often shocks human observers. Indeed, when domestic bitches start fighting in a family home, some of the most experienced behaviorists accept that they can do little but advise them to be separated permanently.

Wanderlust

Dogs are a social species, so the benefits of communal living usually outweigh the risks and disadvantages of leaving the pack. Clearly, for domestic rather than feral or wild dogs, the choice to stay or disperse from mother, littermates and, indeed, their group of humans is limited by the constraints of collars, leashes and fences. But why do feral dogs leave their family groups? Do they jump or are they pushed? Dogs tend to leave their family because the lures of socialization and sex are stronger than sticking to the familiar. Rather than aspiring towards a two-acre block with picket-fencing and a family, they see playmates straying and wander off with them, free to discover resources and exploit the group for as long as possible.

Coming back to the family group

What happens should feral dogs, having once dispersed, return to their family group? They may or may not be accepted. Either way, they improve their chances of being accepted and fitting in if they are deferent and signal submission or avoid close contact until they are sure how the land lies.

If dogs base their relationships with humans on the relationships they have with other dogs, this apparent dismissive approach to the social group may explain why, when domestic dogs are reunited with their human family after a two-week holiday, they can be a little aloof and distant rather than as friendly as normal. Often interpreted as a show of anger or disgust at having been left behind, this response may simply be the dog's instinctive way of avoiding clashes after a period of separation. After all, in the dog's absence, who can say what may have changed? A new social order could have been established, and so it would pay to avoid affirming strong alliances just in case there's been a changing of the guard.

Accommodating the new member of the family

When new pups arrive on the scene, adults—whether they are pack members or non-pack members—will either play with them enthusiastically or barely tolerate them. Bitches other than the mother are potentially very dangerous to a newborn litter. Breeders often report that one bitch (with or without a litter of her own) sniffs out, stalks and kills pups left unattended by their mother. Infanticide is not a behavior seen in male dogs. Unlike a new lion in the pride that kills the offspring of the previous leader after taking over, male dogs are not known to kill pups sired by another male.

It's worth considering how the fathers of pups may recognize a relationship or, more likely, a familiarity. Although there isn't much research to back this up, it may be that some males may recognize their kin because they smell similar, in the same way that human fathers are said to subliminally recognize facial features in their newborn children. Indeed, it is said that children of both sexes look more like their fathers than their mothers in very early life as a mechanism to avoid suggestions of their being unrelated due to their mother having conceived elsewhere. Perhaps pups may smell more like their father than their mother.

Do dogs treat humans as other dogs?

Dogs never put others on leashes, or clip their toenails, or feed them from bowls. So there are significant limits on any dog's innate ability to relate our behavior to the canine social repertoire. How many dogs march up to an owner eating a sandwich and shove away the human in the same way as one dog might displace another dog? Almost none. So, it seems likely that if we believe dogs regard us as they would other dogs, then we have got only half the story right. Perhaps this is why dog–dog rank does not translate word-for-word into dog–human relationships. And why would it? After all, we behave and communicate so differently from dogs: We walk around on two legs; we eat, hit and stroke with our hands; we bare our teeth in a smile when we are happy; and we don't use our scent in marking, preferring instead to put all our pee and poo in a large bowl of perfectly acceptable drinking water.

Do dogs have a concept of leadership? If they do, we should model our behavior on the most benign leaders whose rank is beyond dispute. Arguably, the most benign leader they encounter is their mother. She led them, fed them, licked their bums and—yes—

disciplined them when they overstepped the mark. This is not to say for one moment that humans should hit dogs. Anyone who wants to be a leader in the true canine sense might have to lick all necessary body-parts of their pack before they mete out physical discipline. At the risk of saying something really obvious, dogs are the only species that can communicate with dogs as well as dogs can—hence the clarity with which they can guide each other's behavior. Hand-reared pups can be very belligerent and prepossessed, possibly because they lacked a canine caregiver's consistent setting of boundaries. Learning a sense of order/rank/status/hierarchy as early as possible is critical for a species as social as the dog, and that is what littermates help to do. If you want to see what over-indulgence of a dog results in, track down one from a single-pup litter. In my experience, these are notoriously headstrong, assertive animals.

Understanding rank

Dogs value resources and access to them. We've looked at the value system of domestic dogs in chapter 3 and can see that two dogs are rarely equally motivated for various resources. Do dogs have a sense of rank? I believe so. They learn that being displaced from a key resource makes another resource less easy to defend. This, in essence, is what matters to animals within a hierarchy. Can I retain the resource, or can I access it by displacing the other animals?

Why would they have evolved to observe rank? Predictable relationships between familiar group members mean that acquiring and guarding resources is more straightforward and less potentially injurious. These relationships are largely underpinned by deference. Nevertheless, some dogs will bite to defend resources currently in their possession. But—and this seems crucial—dogs rarely attempt to displace humans from resources already in their possession. A clear demarcation between human resources (including food, sleeping areas and toys) and dog resources helps to reduce the risk of such a clash.

One of the problems with rank is that each dog behaves differently when it comes to accessing and defending particular resources. For example, in a given household, one dog can have unchallenged access to a certain bed but be unable to defend food. Does rank change with context? The example above suggests most definitely that it does. The value of resources varies with context. The simplest example is the value of toys in one dog's territory: The owner of the territory will value those toys more highly than will visiting dogs. In this case, "ownership" means familiarity with and investment in a territory and its resources. This is why when newly introduced dogs are left together in a backyard, it is critical that all toys and bones are removed.

Does rank rise with age? Generally, yes. Youngsters can usually be displaced from most resources by their elders. In the case of feral dogs, age usually means that a dog has been resident in a group for the longest. So, there is logic in granting status to senior individuals; they know the ropes, or, in the case of a pack's territory, they know the most productive hunting terrain, the sweetest water, the coolest spot in summer and the most sheltered lairs for winter.

Does rank change when a bitch has offspring? This is a moot point as, in feral contexts, it is most often the alpha bitch in a pack that breeds. The presence of more than one litter is unusual but may mark the ascension of a bitch as she spends more time with the prime males, and in that sense her rank may be advancing at the time of the litter's conception rather than being triggered by its arrival. Bitches are certainly aggressive if they sense that an incomer may pose a threat to their pups, and why wouldn't they be? Bitches occasionally kill one another's litters if they gain access to them. This is part of the strategy that allows one bitch to breed and ensures that the pack will channel resources to her brood.

In the light of this brief introduction to dog–dog relationships, we can begin to explore how, as dog owners, we can best make use of that knowledge to live with dogs ourselves. To do so we may wish to use the popular labels: leaders and alphas. But these are troublesome terms because, as we have seen, the role of leader can be shared and a bad alpha can be tyrannical. It seems that historically we have been hell-bent on forcing humans to mimic roles that only dogs can truly fill, roles that are only relevant when they are living with other dogs, rather than with humans. I fear we may have become bogged down in terms that have an appealing ring to humans but fail to strike at the core of what matters to dogs. After all, dogs bond to us because we are worthwhile as caregivers and companions, not because we are alphas or leaders.

The folly of domination

Back in the 1980s, dog trainers believed that Uncle Wolf had all the answers and that the hierarchies observed in packs of dogs and wolves were identical to the social structure of dog human relationships. That was all there was to it. The premise was that if we simply did what some wolves do to each other, we would be communicating with our dogs harmoniously. Even though, as we have seen, most dogs defer to humans all the time, this assumption led trainers and vets to insist that humans should actively step into the role of an alpha, so that, for example, dogs were less likely to bite when defending their resources. However, very sadly, this also gave way to hands-on interventions such as the alpha roll. This involved the human standing over the dog and staring, menacing and even grappling with the dog until it rolled over, as if in submission. The technique did a lot of damage to perfectly healthy human–dog bonds.

The crudity of the signals used by the humans was matched by the variety of attempts dogs made to appease the humans. Sensibly, some dogs squirmed, struggled and resisted this practice. This was often met with an escalation of force from the humans: dogs were pinned to the ground and sometimes defended themselves by attempting to bite. When this didn't work or when it led to more force, dogs urinated or defecated in fear. They were really in crisis. And all the while, well-meaning instructors stood behind the naïve owners and taught them how to perform this abuse. After way too long, the gentle world of positive dog training resisted this approach and effectively rejected the hands-on imposition of rank. Hands-off training was born, and dogs everywhere began to feel the benefit.

Leadership without domination

The emphasis in modern dog training shifted to providing a leader-ship role without domination, coercion or force. This is gentle, humane and effective. It is also great news for dog welfare because it stops owners entering into combat with their dogs. Some top dogs never so much as growl at their group members—they have only to look at them with a fixed stare to discipline the other dog or to be offered a gesture of submission: We should model our behavior on these dogs. Although dogs know the difference between a dog and a human, the absence of any physical or even visual challenge can place us as clear leaders. So, the idea of leadership replaced status.

Are there differences in leadership?

You may well ask: What is the difference between a leader and the highest-ranking member of a pack? Often there is little difference. The effective leader in a pack gets the best of the resources, but so does the highest-ranking pack member. Both can displace others for access to food, water, sleeping places, preferred pack-mates and so on. Both add value to possessions by playing with them. Both endorse the group's activities and interests. Unless there is confusion about rank, due to a change in the group's composition, for instance, the highest-ranking member of a pack is rarely, if ever, actively aggressive. It uses the minimal necessary force to achieve its goals. The parallels seem remarkable. Is there any essential difference between an alpha and a leader?

These days in dog behavior circles, dominance is referred to as the "D word." The notion of dominance brings with it some emotional baggage and conjures images of domination and dominators. And, as we have seen, this can unfortunately be misinterpreted as an endorsement of the use of force. We don't want dogs to displace us from resources, so we need to have a relationship that avoids conflict and helps us to go unchallenged. We can craft our interactions with dogs to deploy the other D words, deference and displacement, to achieve implied ranking but, most importantly, rank without rancor. Leaders generally initiate activities, while alphas reliably displace lower-ranking group members from resources. A leader who uses force runs the risk of being labeled a bad leader. So, the contexts in which these individuals identify themselves are different. This explains why the pivotal differences between the concepts of benign alphaness (dominance) and leadership revolve around dog welfare.

Dealing with disobedience

Let's test the ideas and see where they take us. Imagine you are faced with a dog that has climbed onto a bed or sofa and is refusing to move. A fixed stare wins you either a cow-eyed look or an apologetic beating of the tail tip. Telling this dog to get off evokes no response, and then you can be sure the trouble is about to start. If pushed or pulled by the collar, the dog begins to growl. In the old thinking, this would be described as a dominant dog. If you back away (defer), the threat to the dog's resources recedes, and so the growling is rewarded, making it more likely next time.

The new school of thought says this dog needs leadership. In contrast, I'd simply say it needs training to know (a) the command to "Get off!" and (b) that defiance or a challenge of that sort never pays off. I don't want to see any more of that growling response, so I need to remove the benefits of it and make the dog pay a cost. Strictly speaking, if this is effective and I make the response less likely in the future, I have punished it, but not with violence or abuse. I advise owners to use the tried-and-trusted method of attaching a leash to the collar of a dog that refuses to get off furniture immediately. One command and, if the dog does not get off, follow through with steady traction. No need for hands on collars, and certainly no need for violence. The dog hears the command and has the choice of either getting off by itself or being steadily dragged off. We will explore much more of this sort of approach to training in the following chapters.

In contrast, those who advocate nothing but gentle leadership advise me that the way to get this dog off the bed is to create a distraction in another room. I suspect my own dogs would find this strange but wouldn't allow it to disturb their enjoyment of the bed or sofa. And so I ask myself, "Would canine leaders bother to create distractions?" "Hierarchies are unimportant" is a message sent by very well-meaning people. But in abandoning the concept of the hierarchy, we are in danger of throwing the social-order baby out with the slightly soiled hands-on-domination bathwater.

Understanding aggression

Understanding the motivation behind dogs showing aggression towards humans can be quite challenging. For example, when a dog bites a child playing ball with it, how can we be sure of its motivation? Did the child grab the ball at the very same moment the dog did and so get bitten by accident? Did the dog bite the child to make it release the ball? Was the dog threatened in some way by the child holding its collar before allowing it to chase the ball? There are too many possibilities.

Compared with other forms of aggression, food-related aggression is fairly easy to explain: defense of the food resource. This problem, reported by about 15 percent of Australian dog owners, was the focus of a recent study of the dog–human bond. We studied the possible causes of food-related aggression by asking dog owners to describe the daily management of their dogs and any unwelcome behaviors. The following were associated with a rise in food-related aggression: dogs that were older when they joined the household; feeding the dog a meal on arriving home; having more than one adult female human or more than one dog in the household.

We can readily explain most of these tendencies. Dogs that were older when they joined the home may have learned to growl, snap or bite humans around food in a previous home; feeding the dog on arrival home is likely to mean that the dog is fed before the humans have been seen to eat (an approach that may imply deference in humans); and having more than one dog in the household may increase competitiveness and therefore resource-guarding. However, we failed to establish why being in a household with more than one adult female human increased the risk of a dog being aggressive around food. Further, we couldn't work out whether arousal, frustration and exercise prior to feeding on arrival home made any difference to the dog's behavior. Clearly, more research of this sort is needed. It would be especially helpful to follow a very large group of companion dogs from puppyhood through life to identify the risk factors for various unwelcome behaviors, including aggression.

Dogs given treats during training were less likely to show high levels of food-related aggression, whereas those given treats during dinner were more likely. Although a cause-and-effect relationship has not been identified, treat-giving during human meals may prompt dogs to regard humans as group members who often relinquish food, so the remedy seems obvious. Intriguingly, food-related aggression was often found in dogs that were also reported as having separation-related distress.

Understanding separation-related distress

We imagine that the distress dogs show when their owners leave them alone is similar to that shown by human infants separated from their mothers. This is explained by a theory that says mothers and owners are "attachment figures" that infants and dogs, respectively, cannot cope without. Separation-related distress is important, because some studies suggest it affects nearly half of the companion dog population, and its prevention is poorly understood. It seems worth asking whether, for some dogs, humans represent the best of all opportunities. In that case, some dogs may be missing their resources, not their moms.

Providing consistent companionship

With no entirely benign leader, apart from their mother, why would dogs have evolved to appreciate a benign [human] leader, or to recognize one, or to predict that they can rely on one? Dogs as leaders do not feed one another, so, again, the notion of a straightforward translation from dog–dog to dog–human falls over. It seems that as opportunists, dogs would certainly stick around when they are onto a good thing. A calm leader may be valuable as a consistent, humane provider of its physical and behavioral needs. So, if we look at it from the dog's perspective, being consistent as a life-coach and companion may be more important than being either a leader or an alpha. I teach the principles of good training to all my veterinary students because I believe they need to get this right before they start toying with labels that define the perfect relationship dogs should have with their humans or humans should have with their dogs.

Enriching our dogs' lives

The best owners help dogs to get the best out of every opportunity to travel through life with minimal frustration and conflict. And they benefit by having the happiest, most successful dogs. So perhaps we might do better with and for our dogs if we thought of ourselves less as alpha "dogs" or leaders and more as life-coaches, providing opportunities for them to learn how to have fun and generally get the best out of life, yet all the while actually doing what we want them to do.

Of course, the concepts of leaders and alphas have some merit, but the fundamental flaw is the latest evidence showing that dogs and humans have coevolved. So, while the rules of the dog world certainly operate for Feral Cheryl, they never apply 100 percent when *Canis familiaris* is exploiting *Homo sapiens* and vice versa. We do not have to fit into the dog's world. Though it saddens me to admit it, it is the humans who are in charge of who gets neglected, neutered and euthanized, and so it is the dogs that must fit into the human world first and foremost. Rather than acting as leaders or alphas for our dogs, we need to acknowledge our coevolution. I believe our role as their life-coaches is a more honest, exciting and unique role than any attempt to be a pseudo-dog could ever be. The rest of this book identifies the characteristics of the best life-coaches for dogs.

Playing the networking game

As the Peter Pans of the animal kingdom and in contrast to Uncle Wolf, dogs play well into their old age. If the extraordinary emphasis on play is a characteristic of *Canis familiaris*, we'd do well to study it. Aside from any other benefit, it may help us understand our dog's needs and ensure that we meet them.

Dogs have different play styles for different playmates. For example, one of my own dogs, Wally, knows he cannot rely on small dogs for an impressive, stimulating and challenging game, so he approaches them with a consistently dismissive demeanor and, surprise, surprise, very few little dogs make playful overtures toward him. We have all spotted such prejudices in our dogs. Despite a lack of scientific evidence, I am strongly suspicious that some dogs develop breed discrimination. Dogs learn that others of a certain breed are generally excellent playmates, while those of another breed are not as attractive. It is worth considering how persistent they can be as a result of what one might call a self-fulfilling prophecy. We can set up successful networks for dogs by bearing these preferences and prejudices in mind when introducing dogs to playmates.

INTRODUCING DOGS TO ONE ANOTHER

Planning an initial meeting between two dogs is far from complicated. Fighting between normal dogs and normal bitches is extremely rare. Opposites usually attract, largely because dogs have no idea they are neutered and always retain a residual interest in the possibility of making puppies. Having said that, safety is paramount. If aggression spills over to the extent that the dogs actually begin to fight, separate them and call it a day.

Territory that is neutral for both dogs is very important indeed, but it may be difficult—though rarely impossible—to locate. Fenced areas are best, but they must be large fenced areas. Finding a space that is not occupied by other dogs is pivotal. The emphasis must be on the meeting between the two dogs and on them being able to concentrate on each other. A third dog can form alliances with one or other of the dogs being introduced, and these alliances can even represent a resource worth guarding: an outcome that can spill over into aggression.

Ensure that the dogs aren't distracted by removing anything that could be possessed. The idea is to keep the focus on the dogs' fun and not on anything that can be perceived as a toy. Always keep the resources to a minimum. Remember, the dogs are supposed to engage with each other as they explore the novel environment together, not with you, not with the other dog's owner and certainly not with any toys, balls or sticks that may be around. Equally, it is best to keep walking rather than standing still and chatting or staring at the dogs. This avoids either dog setting up base-camp beside the humans and guarding them.

Without good planning, introducing dogs to each other can be fraught with the unanticipated; dogs can get on famously, as if they have always known each other, or they can race towards each other like rockets only to end up sailing straight on past in different directions as they sniff something far more engaging. If the dogs do not get along instantly or ignore each other completely, they are probably spending time assessing each other.

Playmates can become a valued resource for most normal dogs. The owner should be the keeper of this resource. By this I mean that you should be able to get your dog to sit or offer a trained behavior before taking your dog off the leash or opening the door to a playmate. Similarly it pays to call the dog back to you when it is playing with other dogs, then clip on the leash, walk a few steps away from the other dogs, reward your dog and release it. This dilutes the forbidding, walk-ending properties of the leash and avoids your dog learning to dodge you when the leash is evident. Practicing emergency recall (getting your dog to come to you in an emergency) is well worth the time and effort. In chapter 12 (Fine-Tuning) we look at ways of training a rock-solid recall but I mention it here because it is important to test and develop a good recall response in the presence of other dogs. This avoids your dog becoming one of thousands that respond beautifully to many commands but only when there are no other dogs or external distractions around.

Enemies and alliances

The way dogs relate to each other depends on their breed, experience and the context in which they meet.

OUT AND ABOUT

Although many dogs enjoy playing with other dogs throughout life, a significant number do not. As they age, they develop prejudices, aches, pains and learned play styles that may not gel well with other dogs. Clearly, these are the dogs that should be kept out of off-leash dog parks. Interestingly, large numbers of urban dogs with these tendencies are exercised at night, because their owners have figured out that this is generally less stressful, as it means they are less likely to bump into other dogs. Of course, on a grand scale, this does mean that dogs you encounter after dark may be more likely to be aggressive to other dogs. My advice is, until you are given reason to think otherwise, it is safest to assume that owners exercising their dogs under cover of darkness are doing so for this reason. Steer clear.

PLAY PARTNERS

Unless bad experiences or old age get in the way, every normal dog has a playmate out there somewhere. They learn how to play with each other. The nuances of each dog's game with another dog are subtly different. With one playmate your dog may vocalize, with another he may constantly play-bow, while with others he may chase. The longer dogs play together the more likely they are to develop sophisticated games and evenly disperse their wins and losses.

We know that, despite their undoubted need for company, most dogs live in single-dog households. This speaks of widespread species isolation and probably contributes to many of the problem behaviors that dogs adopt in a bid to cope with the pressures of modern life with humans. However, our survey on separation-related distress showed that the number of dogs in the household had no bearing on the risk of separation-related distress, a finding that's consistent with previous studies. So, getting a second dog as a therapeutic step will not necessarily reduce the problem. That said, don't rule out providing your dog with a playmate, as play with other dogs is invigorating—like jogging for a human. Play with *familiar* dogs is extremely exciting because participants learn one another's play styles. In some ways having a regular doggy visitor is better than getting another dog because the visitor retains his novelty value. By inducing fatigue this strategy can reduce the signs of separation-related distress.

CHOICE CUTS

- The transition to a new home brings a range of challenges for young pups.

- Sibling rivalry among dogs can be serious.

- Dogs communicate in intricate ways—deference by subordinates, mutual grooming, face-licking and barking are just some of them.

- Dogs unless restrained tend to wander, but the benefits for dogs of communal living usually outweigh the risks and disadvantages of leaving the pack.

- Dogs have different play styles for different playmates.

- The value of resources varies with context.

- It is always important to avoid dogs ever learning to use their teeth on humans.

- We should never use hands-on methods to displace dogs.

- Misinterpretations of the concept of social order have created numerous welfare problems for dogs.

- Deference generally declines with age.

- Separation-related distress is common among dogs; a regular doggy visitor is often a better solution than getting a second dog.

- Excelling as a life-coach to your dog may be far more helpful than attempting to be a dog-like superior.

Pups learn social skills from their littermates and their mother.

Elevated resting platforms offer attractive vantage points.
This may explain why subordinate group members may be
displaced from preferred resting spots.

6
Sex, Disease and Aging

Sex

MALES AND FEMALES—HOW DIFFERENT ARE THEY?

When one mature dog meets another for the first time, the immediate priority seems to be to establish whether they are dealing with friend or foe. What should I do with this: fight with it or flirt with it? Both sexes have the basic neural circuitry for typical behavioral patterns, and this accounts for the similarities in the way, for instance, they eat, play and sleep. Differences in their brains depend on the extent to which the masculine or feminine system is currently activated. Even before puberty (before biologically significant concentrations of sex hormones begin to circulate within their bodies), dogs are subtly primed to behave as males or females. So even though they're not focused on reproduction, male pups behave differently from female pups.

CHEW ON THIS

Some bitches display more male-sex-typical behavior than others. Some vets have proposed that the "butchness" of any given bitch depends on the composition of the litter she was born with. Because it takes place in the womb, this is called an intrauterine effect. It's not clear how the company a bitch pup keeps for the 63 days of gestation affects her behavior for life, but there are two possibilities. One is that when a bitch pup develops *in utero* between two male pups, she is partially masculinized by male sex hormones (androgens) diffusing through the membranes that divide one fetus from the next. The other focuses on blood flow, which in the uterus is directed from back to front (from the cervix forward toward the ovaries) and thus may allow androgens from male pups near the cervix to influence bitch pups farther forward.

Naïvely, the uninitiated regard male dogs as the only leg-cockers of the dog world (I partially counter this in chapter 2 [The Challenges for the Modern Dog]) but the particular behavior of male dogs doesn't stop there. According to a 1985 survey conducted by dog behavior experts in the United States, there are many behaviors that are more prevalent in male dogs than in female, including aggression and lack of trainability. Males were more pushy with their owners, more aggressive to other dogs, more destructive, more playful and more likely to snap at children. Females scored better on the above and were easier to housetrain and train for obedience. Indeed, according to this study, the only drawback to owning bitches was their tendency to demand affection.

So why would anyone want a male dog? Well, the answer represents a combination of tradition, prejudices and preferences. Some men have always had male dogs and argue that they are tougher; other owners feel they are getting more of a dog's dog and that they can see the world through a male dog's eyes more clearly because males somehow behave more like the raw dog. The fact that obedience competitions are traditionally separated into classes for dogs and bitches means that dog enthusiasts have accepted that there are core differences and that either sex may excel in different aspects of an obedience trial. That said, interpretations in the 1985 study may be rather dated. With the growing emphasis in pet dog training circles on channeling a dog's willingness to play, higher scores for male dogs for playfulness may translate as a more trainable companion.

WHICH DOGS ARE EASIER TO TRAIN?

Although we can be influenced by our preconceptions, it's always worth listening to anecdotal accounts from dog trainers. Some reckon male dogs are more dependable in the obedience ring than females. Although not always advantageous, male dogs are also often described as much more emotionally dependent on their handlers, especially if they have women handlers. Bitches in obedience trials are also said to be more intuitive and more aware of their handler's mood, picking up their handler's anxieties and sometimes allowing these to negatively affect performance. Having had a young bitch leave my side during an agility trial and run into the ladies toilet, I can relate to this!

Bitches are also sometimes described as being too smart for their own handler's good, consistently picking up new behaviors faster than males and faster than their owners ever anticipate. While this is impressive and merits scientific scrutiny, perhaps it's a little exaggerated and begs the question, "How smart are the owners?" I'm reminded of a notation we sometimes used on record cards in small animal practice: D.M.I.T.O. (Dog More Intelligent Than Owner).

THE BITCH IN HEAT

Unless they are Basenjis or have changed hemispheres, bitches come into season (estrus) twice a year for roughly three weeks. During proestrus, which occurs just before estrus itself, the bitch will be more playful to the male but will bark and growl and not allow him to mate. You might think of this as the advertising campaign, letting male dogs in the neighborhood know that they should get ready to party. As the season progresses, bitches tend to urinate more and, in so doing, spread the message (through odor) that they're ready. The urine from an estrous bitch is more attractive to a dog than vaginal secretions. Later in the season, from days 9–12, the bitch will actively seek out male company, court her suitors and passively relax her tail so that it conveniently lies to one side of the vulva. This and her rather provocative behavior mark her out as receptive—the proverbial bitch in heat.

Dog sperm can fertilize eggs up to 6 days after insemination, and so after copulation, successful mating depends mainly on ovulation. Most bitches accept the approach of the male and will stand to breed several days before ovulation. This means that, although the season lasts around three weeks, breeding on only three or four of those days will result in fertilization. And although some breeders religiously mate their dogs on days 10 and 13, for example, there really is no magic day—surprisingly few of the bitches seem to have read the textbook. The variable timing of ovulation means that one bitch may ovulate on day 8 whereas another may wait until day 32. Unless the bitch is in the same location as the stud and she can be observed for her receptiveness, this variability makes breeding on the right day rather difficult. The only truly accurate way around this is to take regular blood samples from the bitch to check the hormone surge that accompanies ovulation.

Every good Feral Cheryl is promiscuous. She tends to mate with more than one male, which means that in the free-ranging state, litters usually have many fathers. This may be beneficial for her genes by increasing genetic diversity. Young adult males copulate more successfully than older ones. In contrast, the very first proestrus and estrus of a bitch is shorter, and the concentrations of critical hormones are relatively low. This explains why males are more attracted to the second or later estrous periods in bitches.

The breeding female is usually more aggressive to other females in her social group during her season. Given that bitches so often cycle together, we believe this is an attempt to reduce or inhibit the mating of other females.

FOUR STATES IN THE BITCH'S CYCLE

- Proestrus: The bitch attracts males, has a bloody vaginal discharge and a swollen vulva. Proestrus lasts approximately nine days, during which the female will not allow copulation.

- Estrus: During this period, which also lasts approximately nine days, the bitch will accept the male. Ovulation varies tremendously but usually occurs in the first 48 hours of true estrus.

- Diestrus: During this period the reproductive tract is under the control of the pregnancy hormone (progesterone), whether or not the bitch becomes pregnant. It lasts for 60 to 90 days.

- Anestrus: No sexual activity takes place. Anestrus lasts between three and four months.

False pregnancy, when a bitch shows symptoms of being pregnant although she has not conceived, happens occasionally during diestrus. Mediated by the pregnancy hormone (progesterone), it is signaled by mammary development and, in extreme cases, changes in behavior and lactation. It's been suggested that this behavior is one of the ancestral legacies of Uncle Wolf. While only the alpha female wolf may breed, nonbreeding females in the pack may go through pseudo-pregnancy and produce milk at the same time as the bitch with off-spring. These "aunts" can then serve as nursemaids, which contributes to the survival of the young. The male also plays a part in protecting and caring for the young.

NEUTERING DOGS

Neutering is the norm in most developed countries. Dog welfare enthusiasts believe that the benefits of neutering—less wandering and fewer unwanted puppies—outweigh any disadvantages for indi-vidual dogs. These disadvantages include a higher chance of obesity unless food is reduced after neutering, and a higher risk of bone cancer and prostate disease. By doing a quick testicle count, we can identify castrated dogs at a glance. Bitches are more difficult to pigeonhole visually, although most people can detect a certain stur-diness in spayed bitches as they mature that isn't shared by their entire counterparts. It's fair to assume that dogs and bitches, with little knowledge of the origins and effects of hormones and no surgi-cal expertise, do not know whether they or their doggy friends have been neutered.

Dogs use olfactory clues rather than visual information to estab-lish whether they are dealing with a friend or a foe. From observing their behavior, it seems that, to dogs, neutered adult male dogs smell very much like young male dogs and that spayed adult bitches smell very much like young bitches. Adult males meeting neutered males rarely fight with them, presumably because they don't perceive them as a serious threat. Ultimately, mature male dogs with testicles are most interested in finding bitches in season. Nevertheless, they are also interested in young bitches because they represent future mating partners. For the same reason they are interested in females with the hormonal profile of anestrus—bitches between seasons. This profile is shared by spayed bitches. So dogs can't seem to tell a bitch that is between seasons from one that is spayed and will check both for any hint of sexual readiness.

IS THERE LIFE AFTER CASTRATION?

To a varying extent, there is life after neutering. As I've said, dogs don't and can't know when they've been emasculated. Yes, they know their scrotum is sore and swollen, but they can't know that the missing tissues were their body's main source of testosterone and only chance of fatherhood. They often simply carry on regardless, even though their testosterone concentrations have plummeted. In the case of dogs neutered *after* puberty, this means that they may continue to pursue bitches and even, in the case of sexually experienced dogs, to copulate.

By the same token, anestrous females and those that have been spayed do not discriminate between a mature dog that has been castrated and a youngster. If either makes sexual advances, he can expect to be rebuffed completely. The entire dog fares only slightly better. He may get to first base before being sent packing. This is in stark contrast to the behavior of free-ranging bitches that have been known to fight to the death for access to the preferred male. So, as a result of wholesale neutering, we see much less courting in domestic dogs as opposed to feral populations. Since a great deal of dog–dog aggression revolves around sex, this generally reduces the incidence of dogfights.

CHEW ON THIS

There is recent evidence that men and women handle dogs differently (for example, men restrain their dogs longer in threatening situations) and that dogs can differentiate between male and female humans using cues from their voices and even their faces. Given the importance of sex to dogs, it's a pity that there are so few studies of the relationships between dogs of either sex with humans of either gender. For example, anecdotal reports of male dogs shifting their behavior when female humans in the household menstruate suggest that, in their world of olfaction, dogs are aware of more of our physiological changes than we give them credit for.

SPAYING BITCHES

Some owners have reported that spayed bitches tend to be more aggressive and eat more indiscriminately than their entire (non-spayed) counterparts. But I wonder if the reported increase in aggression is a bit of a generalization. When studied scientifically, only bitches that had been spayed at less than 12 months were more aggressive, and indeed had already shown aggressive behavior. Females spayed after 12 months showed no increase in aggression. One compelling explanation for these tendencies points to the calming effect of progesterone, which circulates in the bitch's system for about two months after each season. Spaying too swiftly after a season causes sudden falls in progesterone when they are supposed to be peaking and so may cause the bitch to become aggressive or irritable.

Neutering may increase an animal's chance of obesity by reducing concentrations of androgens and estrogens that prompt energetic pursuits associated with breeding, such as running around searching for a mate; further, estrogens help moderate appetite, so their removal can prompt over-eating. An animal's tendency to eat more after neutering may also be due to disturbances in insulin and leptin activity and reduced glucose use. So, without their gonads, animals become more interested in food and are less likely to burn it up.

DOES NEUTERING CURB BAD BEHAVIOR IN DOGS?

Castration for behavioral reasons is still a hotly contested issue. For too long, vets have advised castration as a cure-all for problem behaviors, despite there being such a tenuous link between testosterone concentrations and barking, chewing and digging. Yes, it's true that male behaviors are reduced or eliminated by castration, but not all males change their behavior after castration. According to one survey of owners, castration was effective in reducing roaming in 90 percent of dogs, but in only 50-60 percent of dogs did it markedly reduce or eliminate urine marking, mounting and fighting. The ardor of sexual passion makes dogs roam, which also explains cases of gate-breaking, fence-scaling and tunnel-digging. Aside from roaming, castration seems less effective in moderating behavior in dogs than in cats, and it's hard to predict whether animals will change after castration or not. Experience of sexual activity is a poor guide.

Disease

Disease is worth exploring because the needs of our pets can change significantly if they are unwell. That said, disease often goes unnoticed if it is chronic and creeps up on the dog slowly enough for the dog to adapt, or if it doesn't actually affect the dog's performance as a companion or a worker.

CHEW ON THIS

A recent study from New Zealand has shown that more than 10 percent of guide dogs have myopia. Despite this, the myopic dogs fared as well as dogs with unimpaired vision in standard guide dog performance tests, such as avoiding large objects on the periphery of their visual field.

Although sick dogs may not fear the unknown, including death, as humans do, we should consider what they might know or feel. Even a simple trip to a veterinary clinic can be distressing, as can procedures both to treat and prevent disease. For example, apart from the prick of the needle itself, injected fluids used in such routine procedures as vaccinations may cause a stinging sensation, especially if the pH they need for long-term storage is not neutral.

At a clinic, even the calmest pet may become alarmed by the fear scent of members of the same species. Cats may also be exposed to dogs, owners may be unusually agitated and floors may be worryingly slippery. The vet may palpate painful body parts while performing a physical examination. If the animal has to be left at the clinic (for example, for tests), it has no way of knowing whether it will ever be reunited with its owners.

As a brief case study, let's consider a middle-aged dog diagnosed with osteosarcoma (a type of bone cancer) of the femur. Its quality of life could be affected by the condition, its treatment and its owner's reaction to both. Here we should consider lameness itself; the way the limb is manipulated, flexed and extended to investigate the lameness; time spent at a veterinary clinic; and sedation (as before radiography). The treatment may well involve the pain and discomfort of amputation and chemotherapy. There may also be postoperative pain and the struggle of learning to balance and walk with three legs.

So, in summary, disease can have both physical and psychological consequences for a dog. It's therefore very important to think of ways we might help them get used to medical procedures and thereby reduce their distress when these become necessary. For example, we might gently palpate dogs in the manner of a routine veterinary examination before some meals. Similarly, habituation can include "puppy parties" at clinics and taking dogs (of any age) to veterinary centers during quiet periods for nonclinical purposes (e.g., to visit the clinic for nothing more than a meal in the waiting room). That said, although regular trips to the clinic may allow pets to predict that a visit may be only transient, fundamentally dogs can't know that they're there for only a limited period or even for their own good. So, it's worth considering the benefits of vets making home visits, and it would be good to see the development of psycho-pharmacological support for sick pets when appropriate.

By definition, disease is likely to cause moderate to severe suffering. In the following section we'll briefly look at how various diseases affect dogs and their systems—the respiratory system (distemper), nervous system (brain neoplasia), cardiovascular system (chronic cardiac failure), integument (ectoparasites), musculoskeletal system (fractures), gastrointestinal system (parvovirus disease) and endocrine system (diabetes). Although we'll consider one system at a time, let me emphasize that in real life more than one system may be affected in an individual dog at any one time.

RESPIRATORY SYSTEM

Wheezing, sneezing, snoring, coughing, retching and gagging are all signs of respiratory disease that interrupt an animal's normal breathing and probably cause some discomfort. Distemper is a good example. It's a condition that involves nasal congestion in its acute stage but goes on to cause brain disease and thickening of the nose leather and digital pads (hence the name hardpad) in its chronic form. Apart from causing some sneezing that generally interrupts normal activities, acute distemper makes it hard for the dog to breathe easily and therefore run and, of course, to detect odors. Any major obstacle to a dog's airways causes a slight build-up in carbon dioxide that then causes the blood to become more acidic, a condition known as acidosis, which, in humans at least, is associated with depression. For many observers, a dog that is sniffing and chasing things is a happy dog, a dog with a satisfying quality of life. In contrast, an acidotic, sniffling wretch with a runny beak and general malaise cannot enjoy life to the full: It can't exercise energetically and can't even smell its food.

NERVOUS SYSTEM

Neurological conditions show up with a wide range of signs, from a mild facial spasm (or tic) through to dementia, via a wobbly gait (ataxia). A brain tumor can cause all of these, so let's use this as an example. Behavioral changes in a dog with a brain tumor (cortical neoplasm) can be very gradual and cause its owner considerable dismay. A dog that begins to show aggression towards familiar humans can attract some negative consequences, ranging from avoidance to exclusion from the household. Because dogs are so social, a significant spell of separation can be very insulting to a dog that has had a strong bond with its owner. Owners may also react by hitting these unfortunate dogs, so pain and discomfort are added to the dog's woes. Despite the beliefs of the "treat 'em mean, keep 'em keen" school, physical punishment usually backfires, whereas consistent kindness pays off significantly. When a dog is bewildered, its ability to learn is impaired, so there's little chance that a dog affected by this sort of condition can avoid repeating its unwelcome behavior. As a result, punishments may escalate, at least until the dog has seen a vet. As tumors (and any other space-occupying lesions, such as abscesses) in the brain enlarge, they make the dog's behavior worse, causing disorientation and increasing bewilderment. When ataxia develops, the animal may traumatize itself by knocking, wobbling, flailing and bumping into furniture. The loss of bowel control may cause anger and revulsion in owners and possibly distress in otherwise house-trained dogs.

CARDIOVASCULAR SYSTEM

The ability to circulate fluids is obviously central to physiological functioning, but early onset of disease can also be harmful. An example that illustrates some of the longer-term consequences of cardiovascular disease is canine chronic cardiac failure. As with the dog in our previous example, our patient in this case won't be able to exercise as well and so may be frustrated when unable to chase a thrown ball, especially among his fitter peers. As pulmonary edema develops he may suffer increasingly lengthy coughing bouts on rising and after exertion. Again he may be acidotic and therefore quite possibly depressed. He may be prescribed diuretics, which could interfere with his water balance, bladder filling and urination, and his general routine to the extent that he is forced to urinate in the house and is consequently banished to an outdoor existence for his remaining days.

INTEGUMENT (SKIN IRRITATION)

Dogs normally spend a fair bit of time grooming themselves, but overgrooming can be an important sign of distress or a response to disease. Ectoparasites are the most important cause of skin irritation. The irritation they cause can establish a cycle of allergic responses, self-mutilation and itchy wound healing. Grooming can become a preoccupation and can even compromise the human–dog bond, since itchy animals are generally less appealing. If they develop pus-in-the-skin (pyoderma, which typically gives them a revolting mousy odor), all but the most devoted owners are often repelled.

Although it's not completely clear why, ectoparasites, such as mange mites and, more surprisingly, fleas, become a particular problem in debilitated animals and those with a weakened immune system; some dogs seem to surrender, concluding that they can do nothing to stop the irritation. Some observers may feel that to scratch like a hound-dog is simply to be a hound-dog. But with effective flea control, dogs are much better off and are generally much less agitated, irritable and restless. They are also generally less aggressive.

MUSCULOSKELETAL SYSTEM

Many owners would agree that part of the joy of being a healthy companion dog is the ability to move athletically. We've already considered how disease can affect a dog's love of exercise. But what about when movement becomes not just wearisome but painful? Fractures are an excellent example. Unless stabilized, they cause pain when the animal stands, attempts to bear weight and even when it lies down. Initial veterinary intervention often includes manipulation of the affected limb, causing even more pain. And immobilizing the fractures can bring their own set of problems, from bandages that are too tight and uncomfortable to "unscratchable" itches.

GASTROINTESTINAL SYSTEM

When animals have gastrointestinal disorders they often avoid food, a sure sign that they are unwell. Anyone who has treated a dog with parvovirus disease will agree that in the disease's acute phase it seems to cause serious disability. As well as vomiting, there is the scouring discomfort of repeated attempts to empty the rectum. Dehydration may result, which then causes the sort of lethargy and headaches humans might associate with hangovers. Meanwhile, the loss of electrolytes can cause either acidosis as a result of diarrhea or the reverse—alkalosis—as a result of vomiting. Primarily affecting younger dogs, parvovirus disease may make it difficult to house-train a dog and may negatively affect the human–animal bond. In older, house-trained dogs, sudden loss of bowel control may cause some vexation and possibly even shame.

ENDOCRINE SYSTEM

Hormonal imbalances can affect not just physiology but also mood. Thyroid disease may affect metabolism and temperament but may be over-diagnosed these days. Diabetes is perhaps a better example, not least because it's more common. It can affect a dog's well-being in many of the ways we've already discussed, including their ability to exercise, to compete with other dogs, and to hold their urine over-night. A vet may advise restricting the dog's food intake as a way of controlling diabetes; this can cause considerable frustration for dogs, especially if they are still hungry and watching when humans or other dogs are fed. Although diabetic dogs seem to never get enough water—until their owners understand their condition and provide for their needs—some owners restrict water to avoid having to get up in the middle of the night to let their dog out to urinate. I strongly advise against this, as the consequences of dehydration are many, varied and potentially life-threatening.

OBESITY

Obesity is the most common nutritional disorder in dogs, affecting between 20-40 percent of dogs worldwide. Obesity is extremely important because its incidence is increasing, and it can affect almost all of the body systems we've already discussed. Defined as a condition in which an animal is 15 percent over its optimal weight, obesity has been associated with a number of serious medical conditions that are likely to reduce the quality of a dog's life and its life expectancy. The health consequences of obesity for dogs include cardiovascular, musculoskeletal and metabolic disease. Other conditions linked to obesity include increased irritability and respiratory distress. Not looking their best, diabetic dogs may embarrass their owners and be less fun due to their lack of interest in play and their inability to exercise. In short, obese animals have reduced general well-being that may undermine the human–animal bond by compromising the social and health benefits of dog ownership. Because of its implications for animal health and well-being, obesity is an important condition for owners to avoid and for veterinarians to manage.

Apartment dwelling, inactivity, middle age, being male, neutered, of mixed breeding, and certain dietary factors have been associated with being overweight. As we've discussed above, neutering can reduce a dog's metabolic rate, so veterinarians should be aware that neutering may increase the risk of weight problems in certain breeds and advise owners to reduce feeding or to use low-calorie diets as soon as the dog reaches maturity. Breeds of dog that are at high risk include cocker spaniels, Labrador retrievers, collies, long-haired dachshunds, Shetland sheepdogs, cairn terriers, basset hounds, cavalier King Charles spaniels and beagles.

Dogs fed scraps seem to be more at risk of obesity, as are dogs that were overweight at 9-12 months, these animals being 1.5 times more likely to become overweight adults. Although having dogs in a household decreases the risk of obesity in cats, it's not clear whether dogs in multi-dog households are affected in the same way. Such dogs may have more opportunities to play but may also be subject to peer pressure that can increase their motivation to feed.

Owners may overfeed their dogs for various reasons. Sated dogs may be less likely to wander, scavenge and swallow foreign bodies. As pet food manufacturers endlessly promote the positive aspects of feeding pets, some owners may associate feeding with nurturing and demonstrating love and largesse. Companion animals learn to expect humans to provide food and know how to demand it. Feeding can reduce annoying attention-seeking behaviors in the short term but, of course, giving in to pressure makes the begging more persistent in the future. The incidence of obesity increases with age as both metabolic rate and activity decrease. Except in very old dogs, obesity is more common in females than in males. It's been suggested that obese owners may increase the chance of obesity in their pets, but it's not clear whether this is related to a failure to exercise their dogs, or if there is a relationship between overeating and overfeeding, or perhaps they simply don't recognize obesity in their pets.

The quality of the human–animal bond seems pivotal in treatment, because an interested and committed owner is more likely to adapt the dog's diet and aim to achieve the target weight than an indifferent one. As well as a weight-reduction program, preferably under a vet's supervision, an exercise program is also an important part of treatment.

Always remember that weight-loss programs can compromise the human–animal bond. The tendency for dogs to scavenge and even wander when they receive less food is a serious problem for owners of overweight and obese dogs. Rawhide chews can be useful as meal replacements because they don't radically increase the daily calorie intake and may help owners cope with denying their pets regular foods or feeding them a less palatable diet. Scavenging during exercise can be dealt with by using a muzzle, but this often doesn't work very well because muzzles are associated with attempts to control aggression. The growing use of head collars for dogs may counter this misconception.

One of the most difficult tasks in managing obese pets is to convince the owner tactfully that the dog is overweight. Perhaps owner education could be improved by introducing a standard weight chart on dog food packaging and in owner-education packs. The pet food industry should agree on a standard weight guide, since this approach would be most likely to work if the same guide were used regardless of the brand. And perhaps low-calorie foods should be cheaper than regular foods.

Aging

As they age, dogs seem to take life and themselves more seriously, just as many senior humans do. And as they age still further they can become demented, again rather like humans. Indeed, the dog is considered to be a useful animal model of human dementia. Although veterinarians have been aware of unwelcome behavioral changes in geriatric dogs for some time, changes that are due to underlying neuropathology are rarely investigated. A number of clinical signs and behavioral problems are now recognized as signs of canine cognitive dysfunction (CCD, effectively a type of canine Alzheimer's disease). Rarely reported in dogs less than nine years old, sufferers tend to be less attentive, less active but more restless, can't navigate stairs, tend to wander and are disoriented. Even though these dogs have intact sensory perception, they become lost in their home surroundings, like people who have trouble opening doors they may have used for years.

Especially distressing for owners are disturbances to the dog's sleep/wake cycle, especially when the dog repeatedly wakes them throughout the night, often with persistent barking. Some dogs develop an altered schedule in which they sleep during daytime hours and are active or restless throughout the night. Extensive pacing, which may worsen at night, accompanies this abnormal routine. The restlessness can appear as distressing to the dogs as it is for the owners to observe. Further, there may be accidental soiling, as some pets develop urinary and/or fecal incontinence not attributable to urinary or systemic medical disease.

Our lab's own research into aging and cognitive decline in more than 1,000 senior dogs has shown that changes in locomotion, notably circling and pacing, are the most useful indicators of dementia. Although it's always difficult to distinguish the truly neurological from musculoskeletal disorders in aging animals, many owners report that their dog is much less enthusiastic when greeting them and interacting with them in general. Others note that their dog no longer recognizes its own name or familiar humans. Clearly, this spectrum of behavioral problems is similar to that observed in people with Alzheimer's dementia. Thankfully there are now medications and diets that appear to slow down the progress of many of the signs of CCD, and, after all, there is little evidence that these compromised dogs experience any shame about their regressive behavior. The difference between quality of life and the concept of quality of life seems critical here. As humans, we fear life in a wipe-clean chair more than dogs ever would.

DEATH

The effects of chronic diseases on dog welfare can be profound. They can also accumulate because some or all of the insults described above can occur in one dog at the same time. The speed of decline into decrepitude depends on breed. Great Danes descend rapidly, whereas toy and miniature poodles generally dawdle into their dotage. The tales of dogs carrying themselves off to a quiet spot to die seem to speak of premonition but may reflect discontentment with the usual resting places, a sign of dementia, or the need for peace and quiet. They certainly merit scientific investigation.

Dogs enrich our lives in countless ways. The amount of money that owners are prepared to spend on their pets testifies to the ever-more appreciative regard in which we hold them. And when they die they bequeath a collar, leash and basket to the next dog and are usually one hell of an act to follow. We cry when they die because we grieve for our own devastating loss. But as we honor them with our tears, we should remember to focus on the years of joy they brought us rather than dwelling on their last days.

CHOICE CUTS

- 🐾 There are clear differences in behavior between male and female dogs, even before the onset of puberty.

- 🐾 Bitches come into season (estrus) twice a year for roughly three weeks on each occasion.

- 🐾 Dogs and bitches can't tell whether they or their doggy friends have been neutered.

- 🐾 Neutered animals have a higher risk of obesity.

- 🐾 The jury is still out on whether castration works to resolve unwelcome behaviors in dogs.

- 🐾 A dog of the opposite sex is the best backyard buddy for your dog.

- 🐾 Our pets' needs change significantly if they become unwell.

- 🐾 Nearly half of all domestic dogs in Western countries are chubby.

- 🐾 As they age, dogs seem to take life and themselves more seriously.

- 🐾 Keeping dogs active and playful helps keep their minds and bodies healthy.

Adult dogs play far more than adult wolves, so we do well to play with our dogs and appreciate the value of play so that we can use it indirectly in training. Even at the age of 13 years, Wally's joie de vivre remains irrepressible.

Familiar canine companions are greatly valued by dogs, and harmonious social structures allow domestic dogs of both sexes to work well as a stable group.

7
What Motivates Dogs?

For many years, dog trainers have been referring to play drives and prey drives to describe their dogs' interest in games and tendency to show predatory behavior. But these days it's considered passé to talk about drives, because they bring to mind the over-simplistic notion of a build-up in energy that can be released only by one set of activities. Instead, we are better off referring to attention (*Are you listening to me?*), salience (*Do you care?*), motivation (*Can you be bothered?*) and satiation (*Have you had enough?*). Of these, motivation is probably the most important—just ask any trainer of birds in a free-flight show, birds that can easily flap away into the wide blue yonder. If the birds are not motivated by food (*hungry enough*) to reach the trained destination (*the part of the open-air theater where the food is*), they will be motivated by other resources and exit stage left. Animals make decisions based on their motivation, experience and prediction of the likely consequences of their actions. In this chapter we'll look at ways to develop and harness motivation.

CHEW ON THIS

Some trainers tell me that they can work out what their animals are thinking. These are the people who claim to "reward the thought." But as an ethologist, I study behaviors and maintain that you can only reward behaviors that you can see. "Thinking is a behavior," I hear you say. True, but a completely covert one. There may well be behavioral cues that suggest an animal is processing information (thinking), but until we understand animal cognition more fully and can begin to record images of brain function in unrestrained animals, we need to avoid making incorrect guesses and instead focus on the animal's behavior. By doing so, we can pick up subtle indications that the animal is about to perform the desired response, and we can certainly reward that. Indeed, the best trainers use such movements—that speak of intention—as the starting point in their shaping programs. Reward the first suggestion of a desirable response, and you're more likely to eventually get the complete response.

What's in a name?

Humans use names to identify and address one another and surnames specifically to show the relationship between family members. Feral Cheryl and her pack never use names. They recognize other sorts of labels that they associate with their friends and enemies. These labels can be current, such as the color and shape of the dog running towards them, or retrospective, for example, the smell of the spot where significant others have been lying.

When a breeder gives a litter its pedigree names, she is making a log of the kennel name and often borrowing aspects of the parents' names to create an extravagant tag, usually sprinkled with a few superlatives. Thus we get names such as Kobelco My Heart. Of course, the dogs themselves are blissfully unaware of such hefty titles. When you give your new pup a name you are hoping to capture the character or temperament of the dog you want to own (e.g., Betty). Clearly, the dog is still none the wiser. In contrast, the nickname(s) your dog ends up attracting reflect the character or temperament of the dog you are beginning to know (Botty, Botswana, Betty-Lorraine, Bettina and Badness).

We use names to get the dog's attention, as a prefix to issuing a command. So, what does the name actually mean to the dog? It can mean: *Pay attention to the human because the noises he is about to make may well be relevant to me,* or *Return to the human because she has become relevant to me.* For some unfortunate dogs, they can mean *Stop what you are doing!* and even *It's time to hide!* I am reminded here of Harold, a very lovable pug I knew, who, when called before bedtime, would hide under the kitchen table because the bed was outside and he detested being put out. The unremarkable things about Harold were that, like most pugs, he was lovable and had a horribly convoluted upper airway, meaning that he made a great deal of breathing noise. The *remarkable* thing about Harold was that when called, just prior to lights-out, he would not only hide but would appear to hold his breath as if he was aware of the respiratory racket he was making and the way in which this increased his chances of being sprung. Did this mean he was cognitively self-aware? Possibly—although it could also mean that at some stage he had simply learned that changes in his breathing pattern delayed his banishment.

Do I have your attention?

Generally, owners value their dog's ability to pay attention. For example, a recent study showed strong correlations between owners' rankings of attention, intelligence and obedience. In most domestic contexts, dogs need attention skills to deal with complex social interactions. Similarly, deficits in attention skills may jeopardize dog–human bonds and explain some unwelcome behaviors. Interestingly, this is set against a background of research, notably by Alex Horowitz, with her landmark study of dogs at play. This indicates that dogs adjust their behavior depending on whether they have their playmate's attention. They use attention-getting strategies such as barking, barging or accidentally-on-purpose dropping toys, chiefly when the intended recipient is looking away. This finding suggests that dogs may recognize the importance of eye contact not only with dogs but also with humans. It is supported by strong evidence that dogs look at humans (preferably their owners) when faced with a puzzle. However, recent Japanese studies indicate that, when they respond to our commands, dogs are focusing on our head direction more than our gaze.

How does your dog react when you say *Hello*? The intonation of *Hello*! when you greet a visitor at home is quite different to the tone you use for the dog when greeting him. If you don't believe me, try it out. When you and your dog have a spare moment and no one is around, look away from him and say *Hello*! as if you are saluting a welcome guest. You will notice that he looks for the newcomer. This contrasts sharply with the response he makes when you look in his general direction and say *Hello*! in the regular voice you use to greet him. The tone makes all the difference and can inform the dog of the relevance of what you're saying. It can also tell a dog when you are in a bad mood and when you are worth avoiding. Upbeat tones used when issuing commands are more effective than dull ones. By the same token, it pays never to utter a dog's name in an angry tone because you want only good associations with this word. Adhering to this maxim can make all the difference when calling a dog in an emergency, as when it is in danger.

It may sound trite, but to get a dog's attention you must be relevant to him. A dog that has learned the cues that herald a walk with its owner is reliably excited because he seems to know that the walk is all about him. You have become more relevant to him because you are making resources available. Good handlers get the maximum effort from their dogs effortlessly. They engage with their dogs with tact and strategically placed energy to get their attention. As John Rogerson, the legendary British dog behaviorist, insists, "He who controls the games controls the dogs"—more of this later. Good trainers not only control their dogs but also create situations that allow their dogs to offer the desired response. For example, if they want a dog to pick up a certain item such as a set of keys they remove all other distractions (leashes, shoes, bags, and so on) when first training the response. Regardless of how good they are at making the desired response easy for the dog, even the stealthiest trainers have to put the shaped responses under stimulus control. This means that the dog offers the response every time the command is issued, and only once the command is issued. For this step in the training process, trainers must have their dog's attention.

The average dog in the average home isn't being addressed most of the time and so quickly learns to pay little heed to most human voices. He knows that he has a role in certain circumstances and that this is when humans are most likely to call him. For example, he takes the floor when someone comes to the door, and why wouldn't he take this role seriously? He has evolved to alert the pack to the approach of incoming traffic, and, what's more, he usually gets the pack's undivided attention when he starts barking.

Inside the house the dog is a sitting target for incoming stimuli, and, although he can move around his enclosure, he is at the mercy of the owner making fundamental changes such as leaving or returning home. On a less profound level, the owner continues to have an effect on the dog by simple activities such as switching on lights (that can influence day-length as perceived by the dog), turning off the television (which can be a source of barking dogs that suddenly appear in the living room and leave just as quickly, to be replaced by a human voice delivering a laundry detergent commercial), and producing cooking odors (that so rarely fulfill their promise because they hardly ever become morsels in the dog bowl).

Tapping into the joys of a walk

The sights, sounds and smells on a walk may be more predictable but nonetheless exciting. When off the leash, a dog can choose to approach them or avoid them. A dog on a walk with his owner might think that he has the human's undivided attention and can exercise some autonomy. This is often very different from how things are at home. Think about the way dogs must fit into a regular human home when they are not the chief focus of attention, which is most of the time for most dogs. So part of the fun of a walk relates to control and who is walking whom. A lack of control of one's environment is generally associated with poor welfare. So being in control in these ways allows the dog to enjoy itself as much as it ever can.

Being left alone at home doesn't induce the same lack of control as being at home with the pack. It is far, far worse. Only certain, especially valued members of the pack initiate walks—they are the humans. Without them the house has no walks to offer. What's worse, all those barely hidden resources that the dog knows are there could so easily be plundered by the individuals most likely to know their whereabouts . . . his own social group. New alliances could be formed without the dog being factored in or included. I'm not suggesting that these things consume the minds of home-alone dogs; rather, I am underlining why it shouldn't surprise us that so many dogs do whatever they can to be taken with their pack for a walk.

What we call a walk should more accurately be labeled an-adventure-window-shopping-social-sporting-sex tour. We see so many dogs turning themselves inside out at the prospect of such a daily romp. And why wouldn't they? For dogs, hope springs eternal. Admittedly, there are those exceptionally rare dogs that don't seem to enjoy themselves when out on a leash. These have most probably learned that there is little they can do to exercise choice while on the leash and, as such, are examples of learned helplessness. Some critics have said that some guide dogs seem like this.

Generally, dogs give each walk the benefit of the doubt and set off with absolute optimism, each departure an adventure in the making. They seem convinced that a walk on a rainy day is not such a good idea only when they feel the raindrops on their heads. Small wonder then that they get excited the moment promising hints start being dropped. Slippers are dispensed with, coats are reached for, and keys are rattled—all excellent portents of an imminent door opening. And then, the all-important confirmation: the sound of the leash.

Woe betide all those owners who fail to look at their dogs when these cues are dropped, for they are missing the opportunity to work with a highly motivated dog. Not an easy dog to train so much as an exciting dog to train. Not a steady obedient animal but one that is determined to get its reward. As the keepers of the reward, owners are beautifully placed to wait for exactly the response that they require. But how many of them can bear to waste a moment studying their dogs? After all, they just want the yelping, leaping, squirming and scratching to stop. And so what if the yelping, leaping, squirming and scratching are replaced by concerted pulling on the leash . . . who cares? These are the owners who clip the leash on and do exactly what the dogs are telling them to do: start the bloody walk! And next time the yelping, leaping, squirming and scratching are as bad as ever. And if owners try to prove they are stronger than the dog and able to control it and enforce a sit, the yelping, leaping, squirming and scratching just get worse. As we'll see in chapter 9 (in which we discuss extinction), when rewards are withheld, things almost always get worse before they get better.

USING A CLICKER

We will look at traditional clicker training and its use in combination with food rewards in much more depth in chapter 10 (The Dogs of Opportunity). Briefly, trainers use a clicker device as a secondary reinforcer by training a dog to associate its sound with food.

Pushing down on a backside to enforce a sit hardly ever works in a hyped-up dog anticipating exercise. Leash tension is the same—it simply gives the dog something to push against. Sarah Whitehead, a leading canine behaviorist from Berkshire in the UK, introduced the term "hands-off training" to emphasize the need for guiding and drawing out good responses (true education) rather than forcing our dogs to conform with the hands-on pressure of leashes and the various gadgets that can be clipped to them. The key here is to tap into the motivation and use a clicker to tell the dog it has done the right thing. The reward in this case is the one thing the dog is extremely motivated to get: a walk. So, owners should click and start the walk as soon as they see the slightest improvement over the previous day (calmness, more sitting, less barking). Using the clicker whenever there is improvement in the dog's behavior can help the owner's timing and reduce the dog's frustration because it acts as a promise of a reward to come. In this way the clicker can contain some of the excitement at the prospect of the door opening because it tells the dog that the point of door opening has arrived.

What keeps a dog motivated?

A dog's values can be innate or learned. Innate values dictate the games a breed might prefer to play, which is why most Rottweilers love tug-of-war games and most collies love chase games. Meanwhile, learned values can be the result of previous challenges: A dog that one day feared water may become a rapid convert once it has learned that the fear of being cold and wet is outweighed by the joy of floating.

Success breeds success. So if a dog has managed to get a resource it will inevitably be reinforced for the behavior that brought about the success. On the other hand, disappointment can sometimes seem to demoralize a dog by training it to bow out of games it will most likely lose. Admittedly, defeatism is more common in humans than in dogs, and most dogs have the true Olympic spirit: taking part being more important than actually winning.

CHEW ON THIS

Until recently it was believed that constant loss in competitive games of possession, such as tugs-of-war, might successfully demote dogs that assert themselves during play. Fresh evidence suggests that winning all games of possession does not affect other aspects of dog behavior, such as status seeking. However, one shortcoming of this study was that it looked at only one breed—the golden retriever—a breed that may be innately motivated to keep holding onto articles, regardless of success or failure.

Dogs swiftly learn that sitting can brings food rewards, adoring visual contact, and petting. Sitting is a small price to pay for such reinforcers. Sometimes the motivation is less obvious. Old, lame dogs will still chase balls even though their bodies are telling them to kick back and let the youngsters get to the ball first. The pain of the pursuit is overcome by the thrill of the chase. In the same way, dogs that chase cars may stop running as soon as the car stops. They enjoyed the running more than the grabbing of their prey.

Sometimes engaging in a given behavior can have competing—attractive and aversive—outcomes. Dogs that win fights, kill rats or defend their food successfully—even if that means having to pay a cost in terms of pain (for example, bites to the face)—will all do so more readily in the future. So, even though they had to pay a cost, they were reinforced by the pleasure of retaining the reward or possibly even the challenge itself.

The other motivator that is easily overlooked is the element of surprise. Recent studies have shown that dogs, like human infants, seem to respond to unexpected events by paying more attention to them. Rewards that come as a surprise seem somehow more valuable than those that dribble into a dog's life predictably. Let's face it: Some owners can be dreadfully predictable with their dogs. Some are downright dull. Others can be full of enjoyable surprises, and while we need to strike the balance between being exciting and consistent, it's easy to see why dogs adore owners that inspire them to live life as an adventure. Apart from being a consistent source of food, these owners also make so much magic happen. They can turn on taps, open doors, drive cars and find annoying grass seeds lodged in chest fur. An attentive owner is a highly valued resource—the dog's mobile opportunity shop. We'll look at other ways humans can coach dogs without seeking their companionship when we explore the world of the working dog in chapter 15.

For a dog, novelty is one of the most important attributes of any toy. So finding your dog a playmate who does not necessarily become a permanent fixture is a really great way to keep him entertained. Arranging for a backyard buddy is often a better way of providing play and companionship than buying a second dog of your own.

Changing motivation

Knowing what will excite your dog is to know what makes him tick. Dogs thrive on activity, and truly inspiring humans are fun to be with, not least because they are a source of action. This explains why dogs owned by slobs may appear disloyal when exposed to more active, energetic and therefore interesting folk. We have all seen dogs that come in from a day's activity and seem to switch off completely and sleep on until the cue that reliably marks their departure the next day. They have developed a rhythm that matches their owner's. The team leader dictates what happens and when. This combined with dogs' behavioral flexibility means that, even though they are most active during daylight hours, these dogs are not necessarily victims of the light-dark cycle. If, for example, they are owned by those who hunt at night (yes, I am thinking of those hunters who, some might say unsportingly, use flashlights to dazzle their prey and so rely on pitch darkness to maximize the blinding effect), the dogs will do the same thing but in reverse so that they conserve their energy by sleeping through the day.

Does the change of seasons affect motivation?

Seasonality can certainly affect behavior and motivation. For example, owners of intact bitches are acutely aware of the seasonal onset of estrus, with its associated shift in motivation and priorities. Once domestic bitches are mature, they cycle twice a year, unless they are Basenjis (the ancient African breed that seem to underline their famed primitiveness by cycling with the same frequency as all wild *Canidae*). This side-effect of the process of domestication has enabled dog breeders to select breeding stock more frequently than, say, zoological park curators, who can select from only one litter per year.

CHEW ON THIS

Seasons had a profound impact on the availability of food, warmth and comfort for Uncle Wolf. To a lesser extent, some modern dogs are also affected by the seasons. Any vet will confirm that dogs that are allowed indoors are more likely to shed their hair to some extent throughout the year. We aren't clear what causes this shift from the annual shedding of the winter coat in spring that is seen in outdoor dogs, but it's most likely due to the controlled environment of indoors. The two prime candidates are heating and lighting. Of these, the more likely is lighting, since detecting changes in day-length is how Feral Cheryl's body works out that the seasons are changing. Natural day-length shifts are virtually obliterated by artificial lighting. In other words, artificial lights easily confuse the biological clocks of animals as long as they are bright enough. Breeders of Thoroughbred horses use this phenomenon to bring their mares into season early. Dogs that shed their coats throughout the year provide evidence of exactly the same effect. Most owners find that if they exercise their dogs more, they notice less shedding. This could be either because so many hairs are lost during exercise or because the dogs are being exposed to more natural light and so are able to set their expectations of daylight to match. The artificial light they are exposed to in the evenings (and to a lesser extent in the mornings) hardly registers by comparison.

The rebound effect when motivated behavior is restricted

Abstinence from a normal behavior rarely causes it to disappear from the animal's repertoire. On the contrary, most behaviors can be made more valuable and so more probable by restricting the dog's access to them. A dog that has been confined to a small space may stretch far more than usual, and if he has been wearing a muzzle he may yawn more than ever. The welfare implications of preventing behaviors with an internal motivation can be profound. They involve frustration and distress. In active copers, these manifest as redirected or so-called displacement behaviors such as self-mutilation. In passive copers, they may take the form of apathy and lethargy.

Behaviors that are prompted solely by a particular stimulus are reasonably predictable. For example, sunlight that wakes a dog might make it stretch and yawn. However, many behaviors don't have one simple external trigger—they are caused by a combination of internal changes and external conditions. Behaviors with an internal motivation have welfare implications because they can help us understand what opportunities and resources dogs miss most. Many dogs perform a highly motivated behavior more often once they are free to do so after a period of being prevented from doing so. Ethologists call this a *post-inhibitory rebound*. Behaviors that show a post-inhibitory rebound reflect behavioral needs rather than physiological needs.

A useful example of post-inhibitory rebound comes from car journeys. The period of confinement in the back of a vehicle without much movement is tolerated by most dogs without complaint. However, as soon as the door is opened and they are given chance to stretch their legs, most fit dogs will trot and canter around. They are doing much more than simply stretching their legs; they are meeting their behavioral need for energetic movement. It's very simple to harness the same principle in training. For example, the opportunity to indulge in any normal activity will be more rewarding if the dog hasn't had it for some time.

Revving-up and restraint

Although it might seem mean, reserving rewards until you see an improvement in a particular behavior can be very effective. If you use food in training, it's worth considering how to whet your dog's appetite and make the most of tasty tidbits as rewards. Preparing food in front of your dog and then putting it away for half an hour before a training session will always sharpen the dog's performance.

If you're using play as a reward, tying up a dog and moving out of its reach before playing with its favorite toy, or for that matter any toy without the direct involvement of the dog itself, is enough to get most playful dogs worked up. But it doesn't work for all dogs. Why not? Is it possible that some dogs are innately motivated by nothing whatsoever? I reckon this is impossible without there being physical signs such as stunted growth. A pup that is not even motivated by food is destined to be a miserably undernourished juvenile. Thanks to the basic nutritional standards in the cheap commercial dog foods and the widespread use, efficacy and affordability of worming treatments, there are fewer and fewer such animals around these days.

When clients tell me that their dogs are impossible to motivate, I always challenge them to reconsider their assumption. Admittedly some breeds are trickier to motivate than others. Bassets lumber to mind. They are difficult to get revved-up, but not impossible. So what is going on with the dogs labeled impossible to motivate and therefore to train? Apart from variable levels of innate motivation, some dogs simply learn to demonstrate little motivation because it is either never rewarded or, more likely, that all their immediate needs are readily met by their owners (their servants?).

The opposite is true, therefore: dogs that know they must work for the resources they value are much more biddable and more pleasant to be around than those that have never learned to value anything. "Nothing is for free" has become the maxim of behaviorists worldwide as they advocate a shift in the distribution of power within the dog–human relationship. They ask the owners of all their patients to wait for a desirable behavior before giving the dog any primary reinforcer (see page 157). To do so is to put the owner in charge of the resources and help them understand the games dogs play and how dogs learn. This then promotes a better understanding of training principles. Without spelling it out in so many words, good behaviorists teach owners to apply learning theory. In chapters 9 to 12, we will look at the principles of learning theory, but just because I have mentioned the word "theory," don't panic. I'll make it as easily digestible as possible.

Learning when motivation has dropped

The triggers for feeding and drinking lie in the brain's hypothalamus and are crudely known as "on" and "off" switches. Feeding a dog to bursting point can cause it to temporarily lose its trained responses. Feeding a dog too much dry food can reduce its appetite and knock out some of the crispness and immediacy of its responses. Appetite is linked to the wetness of the mouth. A dog with a dry mouth cannot eat as swiftly as a properly hydrated dog, nor is its motivation to feed as high.

One of the best cues I have in my trainer's toolkit is "All gone!" It signals to the dog that the opportunities for rewards are over for the time being. Following the principle of non-reward (see chapter 4, where we discuss training discs), it stops them offering responses to get me to throw the ball or pet them or give them a treat. The secret is, as ever, to be consistent. Once you have issued this command, you do not reward spontaneous responses because, in essence, the Op Shop is shut. It is, therefore, a signal for them to relax. Dog trainers should become skilled at identifying when motivation has dropped. Doing so stops them asking for behaviors that the dog isn't likely to offer. This is helpful because associating inferior or inadequate responses with a command can develop a tendency in the dog to offer these questionable responses in place of those previously perfected.

Conflict of interests

A rabbit running out from cover and across a road is enough to send a dog dashing straight into traffic despite the best efforts of the owner to call him back. Competing goals can leave a dog effectively in a dilemma, at a loss to know what to do. This is known as behavioral conflict. Another example is the dog that hates being groomed. It wants to bite the grooming brush or the hand that holds it but doesn't want to be aggressive towards the groomer, so it bites its own tail instead. This is a form of redirected behavior. Dogs in behavioral conflict often experiment with various ways to resolve the competing motivation. The chances of it emerging with a desirable response are slim, so it's best to avoid placing a dog in conflict.

CHOICE CUTS

- Dogs make decisions based on their motivation, experience and prediction of the likely consequences of their actions.

- Knowing what excites your dog is to know what makes him tick.

- To get a dog's attention, you must be relevant to him.

- Upbeat tones used when issuing commands are more effective than dull ones.

- Never utter a dog's name in an angry tone.

- Owners who are in charge of all the resources have the most trainable dogs.

- Learn to tap into your dog's enthusiasm for such pleasurable activities as going for a walk.

- A dog's values can be innate or learned.

- Dogs love surprises; predictable rewards make life boring.

- A period of abstinence can make a behavior more valuable and therefore more probable.

- Dogs that have learned to work for their resources are much more fun to be around than those that have never learned to value anything.

Dogs are the great opportunists of the animal kingdom. Their ability to overcome fear of humans in order to scavenge on our scraps may have been pivotal in their domestication.

Police dogs performing so-called man work are motivated chiefly by the prospect of a tug-of-war game, not by the need to please their handlers or uphold law and order.

8
Bonding with Non-dogs

In this chapter we'll look at the unique relationship that has grown from fundamental similarities between dogs and humans when it comes to mental processing and behavior. Domestic dogs seem to understand human communication more than any other species and so have adapted to our social circumstances.

It is something like a lottery for each dog entering a new home. As the great opportunists, they need to find a human who can help them access opportunities; in other words, they need to be guided to learn the behaviors that will get them the resources. Those dogs that exemplify their breed standard will be chosen by humans who excel in showing them, and these dogs will get the chance to reproduce; the rest of the dog population need life-coaches who excel at getting the most out of *Canis familiaris* as companions. The dogs that score the best life-coaches have the best lives. And those of us who love our dogs want to see them enjoy life to the full; we want to be their best life-coaches.

The role of the human-dog bond

The cement of human-dog bonds is often called trust. This term frequently crops up in books about animals and their training, but what does trust actually mean to Feral Cheryl? How does it define the quality of her relationships with fellow pack members? Your dog trusts that you will scratch his chest when your hand goes towards him; he trusts that you will have food in his bowl when you lower it to the ground; and he trusts that you will throw the ball when you have a ball in your hand. At the same time, he trusts that you will not hit him, offer him an empty bowl or do a dummy throw. Why? Because you usually do scratch him, fill the bowl and throw the ball. Trust is built entirely on consistency.

In training you strike a deal with your dog. Applying a humane set of rules when simply being with your dog rather than when formally training him, maximizes the consistency, which establishes trust, which establishes a bond. This is why bonding can affect how dogs appear to obey rules. Dogs bonded to their owners often wait for permission to begin an activity, while dogs who have not bonded do not. For example, if you compare a true companion dog with an outdoor guard dog, you'll see that a guard dog that lives apart from its owner is virtually autonomous compared with the bonded house dog. The guard dog has no regard for triggers, cues and commands from its owner because he has not spent enough time with him to learn the relevance of that human as a source of opportunities—as a useful coach.

There is clearly a link between bonds and being an effective coach. Verbal praise from a human with whom a dog has no bond is irrelevant to the dog. Most vets recognize that telling a dog during an examination that it is a "Good boy!" is as useless as reading out a shopping list and that they are far better off giving their patients dried liver or a chest scratch. The "Good boy!" has to be learned as a secondary reinforcer for it to be valued, and the human issuing it has to be bonded with the dog receiving it.

Unfamiliar humans may offer worthless praise, but familiar humans can be equally untrustworthy. Owners who tease their dogs often wrongly imagine that their dogs always get the joke. Of course, such ill-advised owners are supremely inconsistent in that, at least some of the time, they have to honor the beneficial deals they have struck with their dogs. If they teased their dogs all the time, the dogs would avoid them or bite them. It is worth considering why some owners are inconsistent, often inadvertently. Some want to show off to their mates or partners. Some are just plain mean. Others attribute too much cognitive power to their dogs and assume that the dogs always know they are only kidding. In canine language the only way a dog can know that a joke is coming up is for the owner to perform a play-bow. This intriguing signal tells the dog that the behaviors that follow are playful or at least not to be taken seriously. Its critical importance is in preventing boisterous playful overtures being mis-read and precipitating fights. Importantly, owners should appreciate that their own play-bow signal should be used consistently as a warning of chase games, stalking and staring, and even rough play.

Dogs love social contact

Dogs have a deep need for contact with a social group. They can detect emotional shifts in their human fellows and will, for example, gaze at them longer when they are watching cheerful movies than when watching sad movies. Social contact is probably the most defining feature of dog behavior. This is hardly a bold claim, but if you don't believe me, consider the following story. I was studying horse behavior a couple of years ago in the Australian bush. I was directed to a herd of brumbies (Australian feral horses) in a state forest one hour's drive off surfaced roads and 45 minutes from the nearest house. I set off with an old friend, and we encountered the herd just where the state park ranger had predicted. We were soon crawling towards them to gather data. To our amazement, we spotted a dog with the horses, a lithe black kelpie cross that was as surprised as we were by the encounter. Far from wanting our company, he nimbly hopped over a fallen tree and fled, forsaking the horses. Given the distance from human activity, we assumed that he, too, was feral and that he returned to the company of the horses when we had gone back to the field station.

It is fascinating to speculate why the dog had bonded with the horses. Maybe he kept close to the herd to scavenge on foal feces (which, being derived from milk, can be nutritious for dogs) or possibly even afterbirth. Another particularly tantalizing possibility is that he relied on the surveillance provided by so many pairs of eyes. Indeed, the horses were the first to spot us as we advanced slowly toward them. Intriguingly, he seemed to respond to the horses' gaze when they spotted us, and only then did he seem to see us and flee, notably well before the horses moved away from us. The likely benefit for this dog of being in a social group seems to have been protection. Unlike regular dogs, he may have had a good reason to enjoy this situation—specifically that he would not have had to share resources with the other members of his group; which, had they been dogs, would have been competitors.

Over-bonding and the risk of separation anxiety

Psychologists speak of attachment and detachment, terms developed chiefly to describe bonds between human infants and their caregivers. The bonds that tie us are dubbed "attachment relationships" by psychologists, a term first used to describe the bond between children and their mothers. Under normal conditions, the process of attachment binds pups to their dams, while detachment effectively severs that maternal–infant bond and makes way for socialization with others. It's thought that an attachment bond between a dog and its human caregiver (the so-called attachment figure) can be artificially created when the human responds to all the dog's attention-seeking demands in a way that its dam never would. One theory is that such a strong bond can lead to dysfunctional hyper-attachment. This over-bonding becomes evident if the dog gets distressed when kept apart from its attachment figure.

The belief that owners can over-bond with their dogs is certainly controversial, mainly because it's a philosophy that can drive an unnecessary wedge between dogs and owners who would otherwise form a completely healthy bond. In fact, according to some studies, dog-related factors were more likely to cause separation-related distress than elements of the owners' behavior. For example, one study showed that dogs whose owners interacted with them anthropomorphically (ascribing human attributes to animals), did not engage in obedience training, and "spoiled" them in certain ways were no more likely to develop unwelcome behaviors than those given more discipline.

CHEW ON THIS

According to a recent study, dogs living with a single adult human were approximately 2.5 times more likely to have separation-related distress than dogs living with two or more people. Low numbers of humans seemed to increase their individual value. The same study showed that neutered dogs were three times more likely to have separation-related distress than entire dogs. With a great increase in the number of dogs living with just one person and the rising (and laudable) tendency for owners to neuter their dogs, it's not surprising that separation-related distress is becoming an enormous problem.

When inappropriate bonds form, owning a dog may become too difficult, or dogs may simply struggle to meet their owners' expectations. The prevalence of separation-related distress is a worry to the pet industry as it tries to encourage more people to own dogs in the face of a worldwide decline in ownership. Few owners seem put off by the potential costs of looking after the physical health of their dogs. If they did, they'd do more prepurchase research to find a puppy that was free from inherited diseases. But if such veterinary costs aren't a deterrent to owning a dog, what is? Well, it seems that many people are put off dog ownership by certain behaviors that they regard as unacceptable.

Perhaps we expect too much from dogs. We expect them to know that we are going to return when we leave them alone, and we expect them to have the social decorum of an acceptable human member of society. The veterinary profession has a vested interest in equipping the public, especially dog owners, with skills that help them to understand how dogs behave and to minimize the most unwelcome—but normal—responses.

Separation-related distress is a good example of how humans can be insensitive to the needs of companion dogs. Indeed, many people laugh at the suggestion that dogs could be prone to such a diagnosis. (They assume it arises when over-pampered pets encounter under-worked vets.) One study in the United Kingdom showed that while 50 percent of dogs had some signs of separation-related distress, fewer than 3 percent were presented to vets for treatment. This may reflect the fact that, although it is the most common cause of complaints to local councils, many owners don't recognize barking as a problem because it mostly happens when they are somewhere else. Further, the study indicates that few owners think of vets as effective physicians when the mind of their dog, rather than the body, is diseased.

Better matching of dogs with owners and their lifestyles may avoid inappropriate bonding, but it's difficult to see how the same dog can be reliably affectionate and yet unmoved when the object of its affection departs. Perhaps the solution is to train dogs to be left alone and even to enjoy their solitude. Separation-related distress is so common that rather than simply expect it, we should develop preventive strategies that reduce the impact of separation from key attachment figures. These include protocols that allow owners to bond with their new pup without over-bonding (for example, the use of an indoor kennel to allow the human to give the dog time and proximity on his terms rather than the dog's). The same approach may be useful for freshly rescued pound dogs, a group known to be at high risk of separation-related distress. The problem is so important that the pet industry should support research to identify management practices in breeding kennels that may reduce the risk of litters developing separation-related distress.

CHEW ON THIS

A recent study of cortisol concentrations (the stress response) in kenneled dogs has shown that levels do not peak until the 17th day of kenneling. The parallels between rescue shelters and boarding kennels are more obvious to dogs than they are to humans: no familiar pack, lots of unfamiliar humans and lots of unfamiliar dogs, mostly barking. This means that we should expect our dogs to be very emotional after only a two-week period in a boarding facility. Essentially, their stress response is likely to be still developing at that stage.

Some bonds are established early

A pup's early experience has a critical effect on its behavior later in life. Innate signaling systems play an especially important role in a pup's early experience, as underlined by recent reports of the effects of Dog Appeasing Pheromone (DAP). Derived from the waxes secreted by skin surrounding a bitch's udder, DAP calms some distressed dogs. As such, it holds promise in the treatment of separation-related distress and noise phobia and is being marketed as a plug-in diffuser. This is particularly helpful because it bypasses the need for even the simplest form of learning—habituation. The owner does not even have to be present for DAP to work. The appeal of such innovations is that we simply cannot make mistakes with it, and it requires minimal effort. This is an exciting possibility, but we should bear in mind that there is no such thing as a quick fix and that any intervention must incorporate ongoing behavior modification. The best dog owners appreciate the subtleties of learning theory even if they do not recognize them as such.

Dogs make us feel safe

An improved sense of security is part of why we own dogs. After getting a dog, many people report that their fear of crime has been reduced. Similarly, owners of hearing dogs report that the dogs fulfilled their main expectation of alerting them to sounds but also made them feel safer.

Encountering new faces

Within seconds of meeting a new face, dogs decide whether they like or dislike it. They work out which party is more likely to be able to move away without seeking permission from the other, and that is

CHEW ON THIS

With some notable exceptions, much of the legacy of Uncle Wolf survives in the behavior of his descendants. The elements of a hunt have been channeled into the various breeds that cater to man's need for canine specialists alongside him when seeking and capturing prey. It's useful to consider the roles that each member of a wolf pack may have taken during a hunt. One can imagine some members of the group being in charge of visual surveillance (the movement monitors) while others were tuned in to changing smells (the scent sniffers). The movement monitors would alert the group to possible prey on the move, especially in their peripheral vision. But cooperative hunting depends on a fine balance of skills within the team: a blend of prey detection (sniffing and spotting) and prehension (grabbing and killing). So, for a team to be effective, the enthusiasm of the visually sensitive pack members had to be supplemented by other types of players. The scent sniffers and the generalists in the pack would have learned to ignore many of the movement monitors' false alarms.

virtually that. Admittedly, the arrival of a third party can swing the balance of power, and there may be future disputes over certain resources, but the status within the pair is usually worked out very quickly. By the same token, dogs seem to work out within an instant or two whether they are always going to be on edge with certain individuals: the rules of getting to know another dog apply to other species. Every first meeting, be it with a dog, horse, cat or a human, tends to influence the success of subsequent encounters with that species. So, if the first meeting is critical for harmony, at least two questions must follow: Should we worry about our body language when we meet dogs for the first time? Yes, by avoiding staring and standing over a dog, we can invite it to approach and assess us. If things have gone badly, can they ever be set right? Definitely. Habituation and delivering rewards are critical avenues for success with cautious dogs. They are discussed in the next three chapters.

Small children are especially challenging to dog owners who want to keep dogs calm and kids safe. Preverbal children are especially tricky and, through no fault of their own, often do all the wrong things. They scream, run, pick up toys and, oh so often, are smeared with traces of chocolate. All of these make them irresistible to playful young dogs, particularly working breeds such as collies. The inclusion of those amazing border collies in the *Babe* movies proved disastrous for the breed because children demanded dogs like the gentle, patient, charming "Fly" from the farmyard flick. Beautiful border collies that should have been on farms working found themselves in urban homes and did what came naturally: they chased . . . anything. In this instance, anything could include children's toys, the little hands that threw those toys, and ultimately the children themselves. And so it was that border collies were surrendered to pounds in droves (in numbers that were matched only by Saint Bernards after the movie *Beethoven*, Neapolitan mastiffs after *Turner & Hooch* and Dalmatians after *101 Dalmatians*).

Language barriers

We smile, but dogs bare their teeth; dogs pin their ears back when they are fearful, but ours are pinned back permanently; dogs have tails, but we do not. These and other anatomical differences make it rather difficult for us to mimic canine signaling with much subtlety. Humans who fall into the trap of assuming that they can speak the language of dogs may fail to recognize the negative effects of some of their behavior and so put dogs under inappropriate pressure. If our body language means anything to dogs, it's either because it approximates something other dogs do or because it's associated with things that other humans have done before. One could spend a great deal of time practicing yawning, lip-licking, displays of deference, and so on, only to use such skills on a dog that has learned to ignore humans because their signals do not compute and are, therefore, virtually irrelevant.

Making oneself attractive to dogs can be a thankless task. You might have seen people trying to befriend an unfamiliar dog, often with a great deal of bending, stooping and crouching, cooing, billing and whining—all with an open mouth and enthusiastic but unnecessary head-nodding, often to no avail. Most dog people know that for every dog that relishes the opportunity to get close and personal with a stranger, there are ten that just walk on by. It is unlikely that we are ever mistaken for other dogs, so we shouldn't expect to be of great interest to dog-focused creatures. We do not walk on four legs and therefore do not match up with the visual signals that tell a dog it may be dealing with another dog. Besides which, we do not smell like other dogs. Even if we have been in the company of sexually mature and intact (entire, not neutered) dogs, we rarely give out enough borrowed odor to prompt either aggression or courtship. Vets are among a small group of people who regularly work with a transient number of dogs and are therefore exposed to the odors of various possible canine contenders. But they are certainly not considered high risk for sexual overtures or male–male aggression. This is evidence enough to debunk the notion that dogs smell us as if we were dogs. Added to this, one has to accept that while crotch-sniffing of humans is a regular doggy greeting and always occurs at the most inopportune moments, bottom-sniffing—a far more common dog–dog activity—is thankfully rare in dog–human interaction!

Dogs are ever optimistic and resourceful

Dogs have short lives, and they seem to value our company very highly, yet often and repeatedly we have to leave them alone for extended periods. We can't really know what their fears are in this situation. The good news is that dogs rarely pine forever. Even if they weep and wail when we first leave, they remain resourceful opportunists. They will happily depart with a new guardian if sufficient enticement is on offer. Grayfriars Bobby, the legendary Edinburgh dog who never left his owner's grave, may be a notable exception, but then again he may have been accidentally trained to stay put by the generosity of those who brought food to him.

Aggression to humans

While separation-related problems are usually top of the list among clients seeking behavioral advice from vets, aggression towards family members sometimes eclipses it. Dogs regularly bite to defend themselves, and veterinarians are among the chief victims. However, dogs most commonly bite when defending their resources. Often the humans around them don't recognize the value dogs place on certain things, which makes it hard to avoid trouble. An obvious example is giving a dog a bone to chew when a toddler is around. More obscure resources include companions and territories (and among the most valued I would include the park that is most often visited and marked).

But despite the undoubted importance of human behavior in dog-bite statistics, there are few strategies in place to prevent dogs with such tendencies being bred in the first place, even though these animals are becoming increasingly unacceptable. Indeed, on the contrary, it has been suggested that qualities such as showmanship, presence and attitude, which are valued in the show ring, may translate on a day-to-day basis into status-seeking, predisposing dogs to fight with other dogs or even defend their resources from humans.

Aggression can be innate or acquired. Breeding for suitable temperament can certainly reduce the prevalence of aggression in dogs. Domestication itself provides ample evidence of this, if any were needed. Dogs that turn on their breeders are routinely culled. However, while we await breeding strategies to create the very best companion dog temperaments, we must acknowledge that the role of learning is also critical. A dog learns to bite if that gets rid of the human hand threatening him. In this case, hand-related threats can include the hand that drags the dog from the sofa, the hand that grooms tangled fur (especially on the tail and hindquarters), the hand that pets in taboo areas (over the nape of the neck of a dog with unchallenged status), and the hand that takes food away. If you've ever removed your hand from a growling, snapping or biting dog— and why wouldn't you?—you have trained it to growl, snap and bite louder, quicker and harder in future. Further, hitting a dog for doing these things is never a good idea because it signals to the dog that combat has commenced. We should train dogs from a very early age to behave appropriately when they have to deal with unavoidable flashpoints, such as going to the vet. This approach is particularly useful for high-risk dogs, such as those bred to use their teeth, as the fighting and herding breeds.

Predictors of aggression reported in recent studies included a few surprises. For example, in one report, female dogs were high risk, particularly small breeds and if neutered. The type of human predisposed to owning a dog that bites has yet to be fully studied. Having said that, a study of aggression in English cocker spaniels showed that owners of less aggressive dogs tended to be older (65+ years) and more attached to their pets. This study also showed that factors thought to cause status-related aggression—such as feeding the dog before the owner eats, a lack of obedience training, and playing competitive games with the dog—did not in fact do so. This is but a tantalizing glimpse into the unfathomed side of the dog and owner pair. We need to consider the role humans play in the emergence of aggression, and indeed all unwelcome behaviors, in dogs.

Being a life-coach rather than a leader

The role of humans in the life of dogs can create confusion or, worse still, conflict with dogs. This is why behaviorists try to provide models explaining how a dog may view its human companions. But the quest for a one-size-fits-all template presents problems. Does the role of leaders in groups of dogs change with context? If so, can we really expect to be leaders to our dogs under all circumstances? Or should we instead accept that the communication between a dog and its canine leader cannot be achieved between dog and human? Again, I'm all for the concept of a coach who helps the dog to exploit the environmental niche of the family home as successfully and agreeably as possible.

Imposing physical pressure or discomfort and so-called dominance over a dog may be seen as applying and withdrawing aversive stimuli and therefore considered a legitimate training method. But as we saw in chapter 5, there is growing distaste for the term "alpha," since this implies domination and permanency. In dog-training circles, this trend has given way to the notion of leadership. But the concept of humans as leaders of dogs brings its own set of problems. It implies that all dogs that "respect" a human as leader and have bonded to him or her will follow that human even when other dogs are around, getting to know each other, sniffing bums and playing. The truth, of course, is that this is more a matter of training and coaching than of a unique dog–human understanding.

Naturally reared dogs (i.e., pups that remain with their mother) will always find those of the same species more relevant than humans as leaders. Perhaps we should simply accept that we are, at best, care-givers and companions, and when we are not giving them care and companionship, we are coaching them. Although there is clearly some overlap between caregiving, companionship and training, there seems to be sense in compartmentalizing them. To do so helps us approach each set of activities with clear expectations.

Some people reckon that just as motivation changes with con-text, so, too, does the eventual winner in any contest over a resource. Let's draw a human analogy: The leader we might select for a trip to the South Pole or to represent us at a professional gathering would be different from one who might lead us to the best bar or club. What we are being asked to do is to nominate the person who gives us the best deal for survival, representation or enjoyment. I'm not convinced that the analogy works, because we have better under-standing than a pack of dogs about the specific challenges and resources of these missions.

There's little evidence to suggest that dogs can make such pre-dictions. Their values seem less fluid than ours. Admittedly, they enjoy fun, but what are the games they play: possession, chasing and killing, and retrieving? Recalling as they do elements of the pack hunt, these are essentially activities that determine survival. Of the three human alternatives, the trip to the South Pole is the best exam-ple of how we might select leaders if we were dogs. We would want clear direction from someone who was consistent and impressive. It would also help if the leader was inspiring, and it would be especially good to know that the leader was innately benign.

How do we behave like a leader? We initiate missions, play, feed and groom, and usually dictate the path taken by the group . . . but not always. In many ways we behave like a most unimpressive leader. We may lag behind on a walk and show no interest in the stimuli a walk offers to a dog. How many humans sniff trees and lampposts let alone mark them? Wouldn't a leader do so? Yet we leave that to other members of the pack. So here again, the notion of owners as life-coaches seems to have more universal merit than that of a leader.

Life on the leash

Heelwork is fundamental to obedience training and can prove invaluable when you want to steer a dog out of trouble and have no leash handy, but being on the leash or walking to heel are not normal situations for a dog. Consider for a moment what this is like for a dog. Being on the leash means:

- permanently invading the personal space of a group member (the human holding the leash)
- risking being trodden on by shod, and therefore potentially pain-causing, feet
- being scarcely able to see the human at the end of the leash, since he or she is facing ahead and busily dodging lampposts.

Dogs do not see the same things as humans. For a start their eyes have different optics that lead to different magnification and visual fields, factors that are compounded by the fact that dogs' eyes are far closer to the ground than are those of the shortest human. Consider for a moment then the plight of a diminutive breed, such as a Yorkshire terrier, as it navigates along a busy footpath. It is, in essence, in a forest of feet and ankles that move about, up, down, forwards and backwards, without much of a predictable pattern. An added challenge for the little dog is that the leash may cause some degree of neck pain unless he can truly predict every last step the owner may take and so avoid any correction/guidance. With these two possibilities, it's not surprising that so many of these dogs are able to persuade their owners to carry them.

Assuming that the handler follows the convention of keeping

the dog to the left, most dogs on leashes are confined to one side of the human, even though intriguing things may be on either side, and town planners never seem to have the sense to put trees on both.

CHEW ON THIS

The long-standing convention that dogs must be led on the left of the handler has its origins partly in the formality of military drill, which demanded that service dogs looked neat and tidy while being put through their paces (in the same way that the sitting trot was adopted by military riding schools to avoid the uncoordinated bobbing and dipping of cavalrymen's heads). But it also harks back to the bad old days when choke chains were accepted as the normal connection between a dog and its human.

A note here for those people who still insist on using choke chains: You must be aware that your dog can never safely cross from one side of you to the other. In doing so, the dog turns the sliding noose in the chain from a quick-release mechanism to a potentially deadly ratchet. Since the unenlightened are the greatest fans of choke chains, the chance of this message being accepted by all choke chain users is bleak. The only option seems to be to ban the hideous things.

Does the convention of leading dogs on the left-hand side have any repercussions for dogs? Some handlers may be following a convention that compromises their handling ability. For example, some humans may be more dextrous with the right hand, while the left is traditionally used for correction and negative reinforcement. So some dogs may be at a disadvantage purely because their handlers are strongly right-handed. Maybe certain dogs cope with the constraints of being on the leash (including neck pain, if choke chains are used) better than others because they are innately left-handed or right-handed. If the left eye is dominant, then they may be more likely to be distracted by visual stimuli, since there is more to see on the left where the handler's leg does not get in the way. This is considered further in chapter 14 on individual differences.

Harnesses are becoming more popular, especially those that direct the head. An abundance of such head collars is now available, each with its claims to superiority. The intrinsic cost of such devices is almost negligible—after all, we are only talking about a couple of strips of nylon webbing and the occasional metal loop and plastic fastener. In fitting head collars correctly (preferably a job for an experienced vet or behaviorist), the critical questions are whether the webbing will get too near the eyes (*if it is safe*) and does the collar stay put (*if it works*).

Other forms of harness are designed for the chest. They can be very useful for dogs that have breathing difficulties or delicate windpipes, such as Yorkshire terriers, which tend to have collapsible tracheas. Dogs that pull on the leash are not necessarily good candidates for these chest harnesses, because harnesses encourage pulling, as in the case of sled dogs. Surprisingly, a harness is thought to be useful in body-building and making a dog stronger. The more muscular breeds, including Staffordshire bull terriers, have attracted an unhelpful fan-base in a sector of the human community with a high tattoo-to-teeth ratio. Here the belief is that harnesses, with clips that allow an increasing number of weights to be carried, will give the dog a workout and beef it up in the process. The evidence for this having any strengthening or even cosmetic effect is slim, but the spirit of machismo that promotes this sort of behavior is alive and well. One wonders what other malpractices from human gymnasia are being perpetrated on these poor animals.

Pet dog trainers and behaviorists often recognize a suite of behaviors associated with the leash. Some call these responses leash frustration. A dog with leash frustration changes its behavior for the worse whenever restrained by the collar and leash. Many are aggressive to other dogs but only when on the leash. Many have acquired this response from being reefed to heel by anxious owners. Dogs with leash frustration often pull on the leash, which means that the owner is pleased to be able to get them off the leash, so they get the dog to an off-leash park as soon as possible. The dog's job is thus to pull them to the point of leash release as quickly as possible . . . and then avoid being recaptured.

CHOICE CUTS

- Domestic dogs seem to understand human behavior better than any other species.

- Trust is built entirely on consistency.

- Dogs bonded to their owners respond to cues; dogs that have not bonded behave autonomously.

- Dogs are supremely social animals.

- Having unrealistic expectations of dogs can put some people off dog ownership.

- Many dog owners don't appreciate the seriousness of separation-related distress.

- Anatomical differences mean that humans can't understand dog language.

- Instead of leadership and dominance, we should aim to be caregivers and companions to our dogs.

- Time spent training a dog to walk to heel is time very well spent.

Dogs do not cuddle one another. During the socialization
period, dogs learn that many of the ways in which we interact
with them are not to be feared.

For some dogs, eye contact with humans can be reinforcing and thus can
perpetuate boisterousness, contrary to the owner's wishes. For other dogs, direct
eye contact can be challenging and can provoke defensive aggression.

9

Learning about the World

In order to understand canine learning (and therefore training), it's important to first clarify our terminology. While researching a previous book about animal training, I discovered that many excellent trainers are confused about the terms used in what is called learning theory. Since one of my chief goals in this chapter is to demystify training, it's crucial that we agree on the meanings of words, especially the technical jargon.

Some helpful definitions

Broadly speaking, any change in the environment that an animal detects by its sensory organs is known as a *stimulus*. A response is any behavior or physiological event. Animals have innate or instinctive responses to stimuli; for example, newborn pups need little help to find a nipple, and young working dogs seek shade with the same intuition they use to herd ducks.

We say that animals have "learned" when there is a relatively permanent change in their response to a stimulus. When they are repeatedly exposed to a stimulus that *reduces* responses (including fear), we call this *habituation*, whereas repetition that *increases* responses (including fear) is known as *sensitization*. As well as seemingly simple sorts of learning, such as habituation, the broad definition of learning includes acquiring knowledge, as in *knowing* that two events tend to go together, *knowing where* important places are located in a territory, or *knowing when* important events occur. Dogs show other kinds of learning that may underlie changes in habits or acquiring skills, as in *learning how* to move a latch so that a door opens or to send a signal to their owner to open the door. Whether any dog ever *knows why* a particular event happened is debatable.

Not all changes in behavior are due to learning. Motivational factors, physiological changes, or fatigue can all affect behavior. A thirsty dog that drinks despite having refused water five hours earlier has not learned anything in the meantime. It's just thirsty. Similarly, when a playful pup is temporarily transformed from a bouncing blur of tongue, tail and ears into a snoozing ball of fluff, fatigue rather than learning is the cause.

In my definition of learning offered above, experience is a prerequisite because the definition excludes behavior linked to maturation. For example, when male puppies progress from squatting to leg cocking, they haven't learned that this new posture elevates the smelly signal they leave for others; they are simply maturing and responding to increased concentrations of testosterone in their bodies.

Changing an animal's behavior: rewards and punishers

Traditionally, there are two ways to change an animal's behavior: using a "carrot" and/or a "stick." Carrots open the way to food, water, sex, play, liberty, sanctuary and companionship. Because they strengthen the responses that lead to these items, they are all effective rewards (primary reinforcers). There are also secondary reinforcers, whose effectiveness depends on their association with primary reinforcers. Meanwhile, the sticks discourage unwelcome behavior, so this defines them as punishers.

CHEW ON THIS

In general, animals prefer things that favor their survival. Less complex animals have a limited range of responses, but all are drawn to attractive stimuli and withdraw from potentially harmful ones. These approaches to and withdrawals from stimuli can often be modified by learning. Invertebrates such as flies, slugs and ants can all learn to modify their behavior. An example of a simple kind of learned behavior in a more complex animal is when a commercial broiler chicken shows a preference for food containing analgesics, presumably because they ease the pain of chronic leg weakness. The reduced pain, though delayed, reinforces their choice of that food over non-medicated alternatives.

Dogs learn despite us

Regardless of our intentions, dogs learn in many different ways. As most vets will confirm, they learn to fear needles very quickly. Indeed, this is why many vets develop injection techniques that spare their patients the sight of the syringe. It's worth considering why any dog would enjoy a visit to a clinic where the vet might manipulate a sore foot to identify the cause of pain, look down an ear with an auriscope, or perform a rectal examination. And even if the vet mutters, "Just relax," in the time-honored tradition of human proctologists everywhere, the dog cannot know how to respond. Apart from the occasional friendly face in the waiting room, there is little to make the visit a positive learning experience. Dogs learn to mistrust the smell of the waiting room, while cats often make themselves scarce as soon as they catch sight of the traveling cage.

Opportunistic animals learn to pick up the slightest cues to the possibility of a reward, even cues that humans provide unwittingly. Seagulls follow trawlers and tugs with slavish devotion because there's a good chance of a free meal. As the arch-opportunists of the animal kingdom, dogs are always on the lookout for food or fun. Some learn that the sound of the car's indicator means that the park is just around the corner. Indeed, some overexcitable (or under-exercised) dogs might then bark every time the driver makes a turn. Similarly, dogs learn to associate the sound of a can opener with supper and come running when any can is opened. Trainers should remember that their dogs may be alerted by subtle stimuli that are inadvertently included in a training program. For example, obedience-trained dogs may associate the forbidding sound of a choke chain with accurate heelwork and become a little reliant on the threat of pain, performing less well when the chain isn't worn.

This goes with that

Almost all forms of training depend on the animal associating particular events. As far back as 350 BC, the Greek philosopher Aristotle suggested that the most important principle of association was contiguity (the proximity of two events to each other). He maintained that the more closely together two events occur, the more likely that the thought of one will lead to the thought of the other. The links don't even have to be logical (as when we memorize poetry or a list of nonsense words that become associated, with each item prompting memory of the next). An association can be based on either spatial or temporal contiguity, or both. For example, a dog may associate his bowl with food by their spatial contiguity and the sound of the doorbell with the arrival of visitors by temporal contiguity.

CHEW ON THIS

The principles of association weren't studied in a systematic way until towards the end of the nineteenth century. Research began with experiments on human memory and then went on to explore how animals make associations. At almost the same time, but quite independently, an already famous Russian physiologist, Ivan Pavlov, and a then-unknown American student of psychology, Edward Thorndike, carried out the first experiments on learning in animals. Despite their different approaches, their research formed the basis of what became known as classical and instrumental conditioning.

Classical conditioning and Pavlov's dogs

Conditioning refers to a type of learning in which the timing of events is particularly important. As we'll see, good timing is almost always critical for dog training to be effective. Studies into conditioning have given us a set of techniques to help modify behavior. In this chapter, we'll explore these studies, and I'll explain how their findings underpin the basic principles of dog training. Conditioning is also a powerful way of studying the nature of associative learning.

Pavlov studied dogs because he was primarily interested in physiology, or how bodies work. However, after winning a Nobel Prize in 1904 for his research on digestion, he became interested in what would later be called the neurophysiology of learning. The change was triggered by something Pavlov and his coworkers initially regarded as an obstacle to their work on digestion. Many of the experiments involved measuring stomach "juices" as a reflection of the kind of food a dog was given. When a dog was tested regularly day after day, a complicating factor emerged: Digestive juice would begin to flow out of a surgically constructed hole in the stomach as soon as the research assistant approached, well before the dog was given any food. The physiological reaction seemed to be triggered by an association between the arrival of the assistant and the delivery of food.

Pavlov and his students eventually realized that this "psychic reflex" would allow them to study how associations are formed. Various stimuli were selected to signal the arrival of food: the sound of a metronome, visual signals, and pressure pads on the dog's body. All of these had the advantage of being easier to control than the appearance of an assistant. Secretions from the salivary glands indicated the strength of the resulting association. The more dribble, the greater the association. Over the next three decades, using this basic procedure, Pavlov laid the foundations for the study of associative learning in animals, becoming more famous for this than for his work on digestion. Although we refer here to the type of learning he studied as *classical conditioning*, many learning theorists still refer to it as *Pavlovian conditioning*.

A typical classical conditioning study starts with a neutral stimulus, one that has little effect on the animal, and presents this *conditioned stimulus* repeatedly, closely followed each time by an *unconditioned stimulus*, such as food. Eventually, the *conditioned stimulus* consistently produces a response, the conditioned response, related to the *unconditioned stimulus*. In Pavlov's experiments, a buzzer (the *conditioned stimulus*) that had little effect when first heard except to make the dog prick up its ears, caused the dog to salivate (the *conditioned response*) after it had been paired many times with meat powder (the *unconditioned stimulus*). If the buzzer was no longer followed by the meat powder, it became progressively less effective in making the dog salivate. Crucially, in classical conditioning a *conditioned stimulus*, such as the sound of a buzzer, is followed by an *unconditioned stimulus*, such as food, regardless of what the animal does when it hears the buzzer. The arrival of food is independent of any response. Thus, classical conditioning lets an animal learn to associate events over which it has no control. Such learning allows your dog to predict events and adapt before they happen.

There are many real-life examples that demonstrate the role of classical conditioning in learning. Some dog breeders make use of a similar effect to ensure reliable performance of stud dogs. They adopt the same routine before taking the dog to the same room prior to every mating to produce conditioned sexual arousal. Even without laying eyes (or perhaps that should be nostrils) on an estrous bitch, the dog is ready for copulation as he enters the room. Similarly, the sound of gravel on the drive tells most dogs that someone is about to turn up at the front door. The sight of a tablet packet spells trouble for some dogs because it is linked with the need to tolerate fumbling fingers down the back of the throat. It is all about associations. So it is the same process that helped Uncle Wolf learn which howl went with which pack member.

Turning a negative into a positive

A particularly useful variant of classical conditioning is called *counter-conditioning*. This is when an unpleasant stimulus is changed into one that is positive for the animal. The first known example came from Pavlov's lab. He used a mild electric shock, which initially elicited signs of pain, as a *conditioned stimulus*. After the shock had been paired repeatedly with food, it started to encourage salivation, and there was no sign that it was still painful. A more familiar example is the humble leather collar and leash. A pup might initially scratch and resist the collar and leash, but very soon it will associate them with the adventure of a walk.

Rewarding behaviors that are mutually exclusive to unwanted responses—for example, in a dog that races after cars, sitting calmly with him near a busy road—is called *counterconditioning*. It can be very useful in behavior therapy and in getting dogs to accept painful therapeutic procedures.

The importance of timing

Some of Pavlov's early experiments confirmed the importance of Aristotle's principle of temporal contiguity. Conditioned salivation in response to a buzzer developed much faster when the food arrived within a few seconds of the buzzer sounding than when the gap between the two events, the interstimulus interval, was longer. However, for training a conditioned response, the optimal length of the gap depends on the hoped-for response. At one extreme is eye-blink conditioning, which can be produced by sounding a beep before delivering a puff of air to the eye. The best interval for this is about half a second. With a delay between the beep and the discomforting puff of more than a second or so, it becomes difficult to get a conditioned blink to the beep. However, the arrival of food to a hungry dog or water to a thirsty one still allows strong conditioning to a light or sound even after a gap of several seconds.

Associations between two events are achieved more quickly if the events are novel. If a dog is exposed to a conditioned stimulus (such as a verbal command) on a number of occasions before the conditioning procedure begins (before it is paired with any training), it will take a while for it to learn the conditioned response. The dog may simply learn to ignore the stimulus because it has no important consequences. This is why some trainers feel it is better to add the cue after the responses it has initiated. So, by way of an example, to train a dog to spin 180 degrees (for freestyle heelwork), I will first lure it with a treat; then give the treat only after it has followed my hand as it describes a circle; then wean it onto a circular finger movement, and finally, when it is spinning merrily with this minor visual cue, I will replace it with a verbal command such as "Round!"

The importance of consistency

Pavlov found that when a *conditioned stimulus* (the sound of a metronome) was paired with food as an *unconditioned stimulus*, it continued to make the dog salivate just as long as the *conditioned stimulus* was followed by food. If the metronome sounded again and again but no food arrived, the dog stopped salivating to the tone, an outcome that is called *extinction*. Extinction applies to all the examples of classical conditioning given here. The dog will eventually stop blinking to the beep if the air puff no longer follows. The stud dog won't get aroused by trips to the breeding barn if none of his sexual encounters ever occur there. Anything that weakens desirable associations can compromise training programs. If I interchange "Round!" with "Go round!," "Now round!" or complicate matters by sometimes adding the dog's name (e.g., "Tinker, Round!"), I should expect a diminished quality of responsiveness, with the dog spinning sometimes but not others. Put simply, inconsistency impedes training. The best life-coaches put great effort into being as consistent as possible.

CHOICE CUTS

- Principles of training apply to all species over a very wide range of conditions.

- Food, water, sex, play, liberty, sanctuary and companionship are all effective primary positive reinforcers because they make the responses that gave rise to them stronger.

- Dogs learn to pick up the slightest cues when there's a possibility of a reward.

- The more closely together two events occur, the more likely will the thought of one lead to the thought of the other.

- Classical conditioning is the same as Pavlovian conditioning.

- In classical conditioning, a conditioned stimulus is followed by an unconditioned stimulus, regardless of what the dog does when it detects either.

- Associations between two events are acquired more rapidly if the events are novel.

- Good timing defines good training.

- Inconsistency impedes training.

Ivan Pavlov studying one of his laboratory dogs. The disc on the side of the dog's face is a fistula used to collect saliva as a measure of the association dogs had between food and various conditioned stimuli, such as noises.

10
The Dogs of Opportunity

While Pavlov's dogs were drooling in Russia, Edward Thorndike was awarded a PhD in psychology in New York in 1898 for a thesis entitled *Animal Intelligence*, reporting the results of experiments on learning in chicks, dogs and cats. Thorndike used mazes to study chicks, but for dogs and cats he constructed a set of what he called "puzzle boxes." The animals had to find a way of opening the box's door to reach food just outside. One animal might need to pull at a loop of cord in a corner and another to flatten a panel on the floor. After the successful response was made and the food eaten, the challenge was offered again some time later. At each first trial the cat or dog took a very long time to stumble on the effective response. Over many subsequent trials the time decreased progressively, but somewhat erratically, until the animal performed the response smoothly and with hardly any delay. Thorndike produced the first learning curves by plotting these times on a graph.

Trial-and-error learning

In Thorndike's puzzle boxes the animal was taught by trial-and-error learning, performing a response (for example, pulling) to get a reward (liberty and food). Unlike Pavlov's classical conditioning procedure, whether and when a reward occurred in a puzzle box depended on the animal's behavior. So, rather than making an environment more predictable, this type of conditioning made it more controllable. Receiving the reward strengthened the correct response. This is known as *reinforcement*. A particular behavior is more likely when it is followed by a reinforcer. If a donkey is given a piece of carrot each time it moves forward, the carrot is a reinforcer. This kind of learning is now generally called instrumental conditioning, since the response is *instrumental* in getting the reinforcer.

By contrast, a carrot dangled in front of the donkey, but always just out of reach, may function as a lure, or bribe. To the extent that this causes the donkey to move forward, it can be seen as an example of *classical conditioning*. In a type of classical conditioning called *sign tracking*, an animal will approach a visual stimulus that promises a positive outcome, such as obtaining food. The donkey probably knows that something that looks like a carrot tastes like a carrot, so offering a carrot is likely to encourage the conditioned response. But this won't continue to be very effective if the sight of the carrot is never followed by an opportunity to eat it; instead, the association will become extinct. Inconsistency impedes learning. This is why enlightened dog trainers hide the rewards in hip packs. They may (or may not) use the food to reward improvements in their dogs' responses, but they avoid having it on display—in their hands, for instance. This avoids the dogs learning to work only when food is visible and thus helps them to generalize their responses and expect to be rewarded even when food isn't around.

The idea that a large range of rewards—not just access to food or water—could be powerful reinforcers was central to the views on instrumental conditioning put forward by the famous American psychologist B. Fred Skinner, who was fascinated by the relationship between responses and reinforcers. In the 1930s, Skinner designed an experimental chamber with three main components:

- something the animal could operate (or manipulate): a manipulandum, usually a lever for a rat or a plastic pecking disc for a pigeon
- a device for delivering reinforcers in the form of small food pellets; and
- some stimulus source, such as a light or loudspeaker that could be used to signal whether a particular response-reinforcer relationship was operating

This new research tool advanced the study of instrumental conditioning and quickly became known as a Skinner Box. Its great advantage over a maze or puzzle box was that the experimenter did not have to intervene each time the animal received reinforcement. Skinner referred to this type of learning as *operant conditioning*, since the behavior "operates" on the animal's environment. Thus, operant conditioning is a type of instrumental conditioning.

Most dog training is based on operant conditioning. It's worth thinking of any dog learning a new response as being in a giant, glorified Skinner Box—the world. To get rewards, Feral Cheryl must recognize "levers" and learn what to do with them. Good coaches simplify learning opportunities so that they stand out as exciting and approachable levers for the dog during trial-and-error training. They shape new responses by rewarding appropriate attempts and improvements.

In a Skinner Box a dog can press a lever at its own pace. Reinforcement is delivered, according to some automatically controlled preset arrangement. The Skinner Box revolutionized the field by making it possible to study the effects of varying the relationship between responses and reinforcement, or various schedules of reinforcement (see chapter 12 [Fine-Tuning]).

Skinner's boxes had a further advantage over Thorndike's. While Thorndike had to wait until the animal made the required response for the first time and only then delivered reinforcement, Skinner could train animals very rapidly if he began by reinforcing early attempts of the target behavior. For example, to train a dog to press a lever, he would start by delivering a food pellet if the dog moved to the appropriate part of the chamber, then when it made close contact with the lever, and so on. This is called shaping, and it can be a way of modifying instinctive responses or of developing some entirely new pattern of behavior.

Shaping for success

Shaping is fundamental to all successful dog-training plans. Although I'll continue to use examples of dogs in Skinner boxes, I must emphasize that in such devices the quality of learning opportunities is unique because there are no distractions. Remember, the lever I refer to in these examples is simply a piece of apparatus that the dog interacts with. Once you are clear about the outcomes of various experiments, you can take the example of the dog pressing the lever and replace it with any other behavior—from fetching a ball to staying (out of sight of the owners as required in obedience trials) for ten minutes. Simply by performing the required response, the dog is interacting with his environment to gain a reward. Even elaborate behavioral chains, such as those shown by animal actors, can be broken down into steps, each being a lever of sorts. Completing the first task in a chain makes the second response appropriate.

Generally speaking, animals can't be trained to perform the final step until they can confidently perform the preceding step. In training terms this is referred to as "reaching criteria". In other words, good trainers understand that the response they want can be achieved if the rewards are *reserved* and they're prepared to wait for it. This is where patience comes in. Day-to-day training usually occurs with a series of small improvements within these steps. That said, seasoned trainers happily acknowledge that opportunities to reinforce exceptional improvements may occur serendipitously. When this happens, they are always prepared to reward. Part of the art of good coaching is to develop a sense for the usual size of the improvements. This depends on targeted responses, the individual dogs and the stages in their training.

Crucially, reserving reinforcement makes dogs try harder. If Feral Cheryl had been fortunate enough to be surrounded by slow rabbits, she might never have learned to break out of a trot. All rabbits are not slow, so she is forced to break into a canter to catch them. Trainers who reinforce only what the dog has already learned will be offered those behaviors and no more. On the other hand, trainers who wait for tremendous strides usually reinforce too rarely and find their dogs lose interest. To train a dog to weave through the poles on an agility course would take a long time using Thorndike's approach. It would be much faster to use a shaping procedure in which the poles are placed well apart at the start and are brought a little closer to one another each time the dog successfully and swiftly navigates a path through them.

Paychecks for dogs

Whether something is called a reinforcer depends on the effect it has. Its merit can be measured in terms of how much it makes the behavior more likely in future. Although words of praise may get a good response from humans, they can have a neutral or even confusing effect with animals. According to the above definition, reinforcement hasn't happened unless the dog's future behavior at heel improves after the trainer says, for example, "Good dog." Instead of encouraging owners to praise their dogs, many enlightened dog schools tell their humans to "make that dog's tail wag"—in other words, to reward the dog with something it values. All of the good things described in chapter 3 can be used as paychecks for dogs.

Sometimes it's not clear to us whether something can act as a reinforcer. So the outcomes of a dog's actions may bring rewards that we aren't aware of. We know that even visual contact can act as a reinforcer (which is why, if an attention-seeking dog jumps up at you, you shouldn't look at the dog while you dislodge him). Often owners rue the day they were demonstrably amused by their dog's inappropriate behavior. A familiar example is chasing a pup with a treasured item of clothing in its mouth. To the dog this seems to confirm that it has found not only something to be prized and held onto but also an effective way of getting their owner's full attention. While dull and muted words of praise barely impact on a dog's future performance, dogs respond to unintended human responses such as laughter and applause as if they were strongly reinforcing. Tail-chasing in pups is a good example of an unwelcome behavior that can be reinforced by human laughter.

Use of rewards

There's a certain strategy for using food, or toys for that matter, in training. If you show a reward to your dog before it performs the response you want, you may be accused of offering a bribe, a term loaded with negative anthropomorphic connotations. I don't advise this because it can train the dog to work only after you've shown it the color of your money. Furthermore, it can tempt some dogs to mug you for the treat. And if you happen not to have any rewards with you on a particular occasion, you're out of luck.

Many trainers avoid using reinforcers in this way, since it can make dogs look closely for what you've got and encourage them to respond only if they see the reward. Hand-held food (or toys) may also distract them and make it less likely that they'll learn novel responses because, while staring devotedly at the snack, they interact less with their environment during a training session. In other words, they cannot think outside the (virtual Skinner) box. When training more than one dog simultaneously it is worth remembering that there is strong evidence that they quickly develop expectations of what rewards they are due and resent unequal reward distribution. Like monkeys, they are described as "rejecting unequal pay." Building cooperation between dogs relies on trainers being demonstrably fair.

In reinforcement training, some owners get confused about what size reward to use. Research suggests that the bigger the better—but not always. A reinforcer may become less effective when many reinforcers are given within a short time: for example, the use of food rewards to a hungry dog. With frequent large rewards, it will soon lose its appetite. For this reason, many trainers use small rewards within a session and end with a large reward, a jackpot, when the dog performs particularly well. Learning theory suggests that jackpots should be used sparingly, because once the dog starts to expect large reinforcers, small ones may start to lose their potency.

My advice is to start training with large rewards and short sessions and then progress towards smaller rewards in longer sessions. Some dogs get over-excited by large rewards, so, as mentioned above, it's better to conceal the jackpots before they are given. This seems to enhance the reinforcing effect of the jackpot, as does making the jackpot novel and tasty (see chapter 3 for what dogs value). We also tend to overlook the dog's joy at discovering how to score a jackpot. The more you can coach your dog as a fellow adventurer and make training sessions joyful voyages of discovery, the better.

Timing of rewards

Instrumental conditioning is as sensitive to the time association between two events (temporal contiguity) as classical conditioning. The shortness of the interval between response and outcome is crucial if instrumental conditioning is to be effective. A dog pressing the lever in its Skinner Box will learn much more slowly if there is a delay of several seconds before a food pellet arrives than if reinforcement is immediate. If the delay between response and reinforcer remains long, then even when the response is well learned, the dog will continue to press its lever at a much slower rate than if reinforcement were immediate.

When first learning to shape a dog's behavior, owners are often slow to reward the behavior after the dog has performed the target response. Within this short delay the dog may have started doing something else. This can mean that the more recent behavior gets the reinforcement, and as a result, it becomes stronger than the target behavior. This can prove very persistent. A simple example can be seen when training a dog to vocalize on command. If you delay the reward too long after the bark you are effectively rewarding silence.

A similar issue arises with delaying punishment. For example, an owner may see his dog scavenging, call him back and hit him. In effect, he is punishing the dog for coming when called, not for chewing garbage. Although there's room for error in both reward-based training and punishment-based training, in the rewards-based system the risk of error is much lower. This is just one of the many reasons I never recommend physical punishment.

Adding value to stimulus and target training

By placing food on or near a lever in a Skinner Box, trainers can tempt animals to that area. This is called *stimulus enhancement.*

Another form of stimulus enhancement may involve trainers or other animals interacting with the stimulus. Playing with a ball before throwing it increases a dog's interest in it. Drawing the animal's attention to a stimulus is also important for a handy technique known as target training. This involves shaping the animal to stay close to a stimulus (the target) that can be moved around by the trainer. While some dog trainers use paddles, rods and sticks as targets, others use their hands. Incidentally, many of us inadvertently use our hands as targets when directing dogs (to get off furniture or into the car).

Repeatedly placing rewards in a certain part of a dog's environment makes it more likely that they'll return to that spot. In training exotic animals this is known as the animal's feeding station, and establishing it is of tremendous importance in keeping trainers safe. On film sets it means animals can be sent to that spot, known as the animal's place or mark, without their trainers being in the shot. Clearly, when a film script requires more than one animal to move in the same scene, each must be taught its mark individually before they rehearse together. We can achieve the same result by using tiny portable devices that emit characteristic sounds that animals are trained to associate with food. On a film set, these devices are small enough to be hidden, and their sounds can be craftily edited out.

Capturing good behavior

A delay between response and reinforcement is not such a problem for effective training if some event fills the gap. If a house light in a Skinner Box is immediately switched on when the dog presses the lever, the dog may learn the lever-press response rapidly and continue to perform it at a high rate, even if the food reward takes a while to arrive. But for this to occur the house light must be a consistently reliable signal for the arrival of food. In this case the house light is said to function as a *secondary*, or conditioned, reinforcer. This contrasts with a *primary*, or unconditioned, reinforcer, something that doesn't depend on special training or experience to establish it as such. All the goodies described in chapter 3 are primary reinforcers. Secondary reinforcers are usually established through classical conditioning. Like any conditioned stimulus, they can become extinct. So "Good boy!" is valued only if it is persistently associated with primary reinforcers such as food or fun. It becomes redundant and irrelevant if not paired with good outcomes, at least from time to time.

CHEW ON THIS

The concept of secondary reinforcers is evident in nature. For example, Uncle Wolf and Feral Cheryl may have learned to associate the smell of rabbits with the thrill of a chase and the meal that sometimes followed. Detection of the smell is then likely to encourage them to persist in their rabbit hunts.

Capturing good behavior relies on excellent timing so that one can effectively label a given response as desirable. We must be quick to reinforce, because what's important for the dog is that the response pays off. So we need a bridge between the performance of the desired behavior and the arrival of the paycheck. This is where a secondary reinforcer really comes into its own.

Secondary reinforcers are most effectively established when given before or up until we present the primary reinforcer, just as for any other kind of classical conditioning. Giving a reward at the same time as a novel secondary stimulus is less likely to work, because the primary reinforcer will overshadow the new stimulus. Similarly, presenting the secondary stimulus *after* the primary reinforcer is unproductive because, although an association will exist between the two, it doesn't help the dog to predict the arrival of goodies. If the door bell sounds only after visitors have entered the house, it can never help to predict their appearance.

Clicker training

The best example of a secondary reinforcer used in training is the sound made by a so-called clicker. This was introduced by Marion and Keller Breland, who studied with Skinner in the late 1930s. Pioneers in animal training, the Brelands developed various feeding devices that made a characteristic sound as a prelude to, and promise of, food. The first step in any of the Brelands' training was to teach the animal to associate this sound with the arrival of food, as shown by the animal reacting with increasing speed and reliability to the sound. The use of a hand-held clicker was the next step. Essentially the click comes to mean: *Yes, that's good. Job done! Expect a reward any second now.* The click forms a bridge between the response and reinforcement (it is also sometimes called a bridging stimulus). When a clicker is first used, the correct association is established by making the sound just before giving a delicious reward. Doing this many times convinces the animal that the signal is reliable. A trainer can be sure that an animal has made the association between the clicker and reinforcement when it stops what it is doing at the sound of a click—and immediately goes to the trainer to receive its reward.

Clickers have revolutionized dog training. Being pocket-sized or attachable to key rings, clickers are convenient but by no means unique. Indeed, as long as they can't be confused with commonly used words, human vocalizations (so-called clicker words, such as a uniquely enthusiastic "Yes!") are even handier. Any kind of signal could serve as a secondary reinforcer. One advantage of commercial clickers is the crisp, distinctive sound they make. The crispness allows even brief responses, such as blinking, to be reinforced.

Experienced trainers know that they must capture behaviors that appear out of the blue, so they are always prepared to reinforce opportunistically and can most easily do so if they make their own secondarily reinforcing sound without having to find a clicker. This said, clicker devices can make praise universal from one trainer to the next (useful when animal actors must perform with relative strangers) because they produce a consistent noise.

One problem for novice clicker trainers is being overcautious in clicking. Delayed clicks allow the animal to proceed to a next step in its response (i.e., to have stopped doing the shaped behavior). Clicker training courses encourage novices to be generous in their clicks, and this in turn encourages animals to be creative and to offer responses readily. Few animals find verbal praise alone terribly exciting. Dogs, especially, prefer edible treats or a game of chase with a ball. That said, under many circumstances using verbal commands and verbal praise is indispensable.

Getting dogs to behave appropriately even when humans aren't around

If delivery of rewards depends on a human being present, the behavior may happen only when humans are around. This means that an animal might not perform the behavior when away from its coach. This is precisely what's going on when dogs pay no heed unless on the leash. The cue of the coach close by has become necessary for these animals to behave themselves. But in place-training and shaping animals to travel from A to B, it is often important to train movements away from the trainer. So how do you tell your dog it is doing well when you're unable to stand beside it and give it food or some other reward?

One technique is to plant primary reinforcers in places where your dog will find them shortly after an improved demonstration of the desired behavior. To train a "send-away" in an obedience dog, the trainer might place a food reward less than a yard from the dog and mark the spot with an even more obvious visual signal, such as a traffic cone. Given the signal "Away!," the dog moves toward the food (i.e., it performs the desired behavior) and is rewarded as soon as it completes the trip to the cone. Then the dog must be returned to the starting point and the food and cone placed farther away. With gradual extension of the trip, the dog learns to rely on the cone as a useful target. Many dogs' enthusiasm for the send-away comes to depend on having first seen the trainer go to the cone target and appear to drop an item of food. It is up to the cunning trainer to break down this association. Enlist the help of a third person to place a reward at the target without visiting it yourself. In fact, many well-trained dogs complete such tasks impeccably without having to rely on such cues because they've already learned to run straight on command.

CHEW ON THIS

In sled-dog training, rewards are also provided away from trainer and den. The adventure and possibility of spotting other animals to chase rewards them every time they negotiate a turn. Similarly, catching sight of potential prey is what teaches racing greyhounds their single most important lesson: to leave the gate as fast as possible.

These days most training of dogs while they are beyond the physical reach of their handlers (distance training) relies on clicker training. Indeed, this is where clicker training really excels. Consistency is extremely important in the early stages of clicker training, when the meaning of the characteristic noise should not be blurred. Keep in mind that the clicker never lies, and never let your dog hear the click without rewarding it, because this can cause some degree of breakdown between the clicker–food association. When using a clicker for training at a distance, it's important to avoid a long delay between the click and the primary reinforcer. Indeed, clickers teach humans to be consistent; they train the trainers.

CHOICE CUTS

- In instrumental conditioning, the response is "instrumental" in obtaining the reward.

- Inconsistency is an obstacle to learning.

- Good coaches make learning opportunities stand out as exciting and approachable "levers" for the dog during trial-and-error training.

- Shaping a dog's behavior is fundamental to all successful dog-training plans.

- Good trainers understand that the response they want can be achieved if rewards are withheld and trainers are prepared to wait.

- It's important to find a balance between rewarding too readily and not often enough.

- Be sure to reward your dog with something it truly values.

- Sometimes we can inadvertently reinforce a dog's unwelcome behavior—such as by laughing at a pup chasing its tail.

- You can add value to a stimulus by, for example, playing with a ball before throwing it.

- 🐾 It's better to conceal rewards until they are given.

- 🐾 Secondary reinforcers function as a bridge between the target behavior and the reward. They are most effective when given just before the primary reward is offered.

- 🐾 The best example of a secondary reinforcer is the sound of the clicker; but any kind of signal can serve as well.

- 🐾 Never let your dog hear the click without giving him a reward.

Face-licking is a feature of puppy behavior thought to trigger regurgitation. Shaping it as a trained response helped timid Tinker become less inclined to automatically roll over whenever she greets humans.

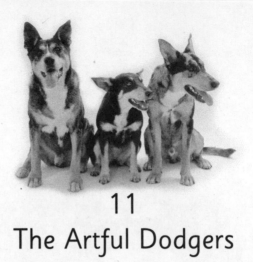

11
The Artful Dodgers

Bad things happen, and all dogs learn to avoid them . . . and, importantly, everything associated with them. A dog's behavior is profoundly influenced by unpleasant, or aversive, events. So, the take-home message is that using a stick—both literally and metaphorically—can produce rapid results. At first glance, imposing punishment or discomfort on a dog in training may seem to work, since it can change some behavior more effectively than positive reinforcement. Yet these outcomes often don't last and have many negative side effects. When it comes to punishment, it's easy to make mistakes. Even our clumsy body language can give a unwelcome message. Take, for example, the way we loom over puppies when greeting them. Is it any surprise that so many pups pee themselves as we stand over them?

Bad experiences (aversions)

Associating training sessions with bad experiences can be highly disruptive for dogs. A single bad experience might overwhelm previously conditioned pleasant associations. So even if we don't repeatedly expose the dogs to horrible experiences, they'll learn to avoid those experiences or their warning signs. Sometimes the warning signs can be the owners themselves, who imposed the discomfort on the dog. Yet the last thing any of us want is for our dog to avoid us. Good preparation before learning opportunities can be really worthwhile if it means avoiding unwelcome discomfort for the dog. For example, assuming your local park has a fenced area, you might want to take your pup off its leash when meeting unfamiliar dogs there. Many pups hurl themselves at all dogs regardless of their familiarity. Others hurl themselves in the opposite direction to avoid unfamiliar dogs. Either way, unfortunate associations may or may not develop. But if neck pain from pulling on the leash *always* accompanies such meetings, the pup may develop antisocial tendencies that persist as it matures.

To explore how dogs respond to unpleasant events and to understand aversions, evasions and phobias, we'll turn again to the literature on learning theory. In behavioral research, scientists may use mild electric shocks, despite their connotations of torture, to discourage unwanted behavior. But before I describe some of the findings of these studies, let me put things into perspective. Most experiments of this kind, mainly with rodents, give the animals a limited number of brief, low-intensity shocks. Scientists prefer such methods because the animal doesn't get used to the shocks. Rats also don't like flashes of light or the smell of a cat, which may generate more fear in a rat than a mild shock, but the effectiveness of such methods drops as the rat becomes accustomed to them. For example, a sudden loud sound that initially produces an extreme reaction may be almost ignored after only a few exposures to it.

Repeated exposure to pain makes dogs withdraw

One way to judge a dog's fear of a certain stimulus—or a warning of such a stimulus—is to test how much it disrupts an established, learned response. When a given response leads to an unpleasant experience, the animal tends to back away not only from that response but also from new responses. This outcome is called *conditioned suppression*. It's important to recognize this dropping-off of responses when there are signals warning of pain or discomfort. In dogs, the same sort of withdrawal might occur if the dog is repeatedly exposed to pain. Studies on shock collars have shown that repeated episodes of pain make significant changes to the behavior of dogs in the places where they have been shocked. The legacies of such abuse include lip-licking, tongue flicking and lifting of the front paw. All of these tell-tale behaviors have been reported to occur reliably in the training arena that have become associated with the shocks. Although similar studies with choke chains have yet to be published, numerous anecdotal reports suggest that the pain of a choke chain can make dogs less and less likely to offer good responses. I hope you can see why choke chains are losing favor.

Studies into conditioned suppression show that fear is learned by the same set of principles as classical conditioning with positive reinforcement, especially temporal contiguity (the time association between two events). For example, the longer the delay between the noise and the shock, the weaker is the animal's fear response to the noise. Essentially, like all other species, dogs will learn to fear something if it reliably heralds an unpleasant experience.

Unfamiliarity sharpens a dog's fear response

Other important principles of fear conditioning resemble those for classical conditioning with positive reinforcement. *Latent inhibition* is when the dog reacts more fearfully if the stimulus paired with the unpleasant experience is new rather than familiar. Your dog's first encounter with a new person can be critical for its future interactions with that person. When meeting a fearful dog for the first time, a human should never insist on interacting with it. The dog is primal with fear—think of her as Feral Cheryl in a domestic dog's clothing. And then think of her being cornered by a human. The last thing she wants is an edible enticement. She is far from motivated to eat, so an offer of food is worthless. She cannot know how well-intentioned the approaching person is or how innocuous the hand is that's extended to her. We've all encountered people who claim to be "dog people" and yet ignore all the dog's stay-away signals; it's unlikely a bond will ever form between these two. It's much better to let dogs approach any novel humans than vice versa.

Generalized fear

Stimulus generalization applies to learning about bad things as much as it applies to learning about good things. Once a dog has learned to fear something, it will show fear of similar things. This is called stimulus generalization; anyone carrying a rolled-up newspaper will become associated with the person who used such a weapon to beat the dog.

Eliminating fear

One final example of a general principle is *extinction*. If a conditioned stimulus previously paired with a shock is presented repeatedly without any shock, it will stop evoking fear. Owners who leave choke chains on dogs when not working them not only risk strangling their dogs but should also expect the sound of the chain alone to be less effective than the combination of the chain-plus-neck-pain.

Most animals confronted by a fearful situation will display the flight response and run away from the source of danger. But by removing itself from the confrontation, the dog loses the opportunity to learn whether this stimulus remains dangerous or not. In clinical psychology and veterinary behavior therapy, specific phobias—such as animal or social phobias—are treated with exposure to the feared object: the needle, the syringe or the vet holding them. Dogs with a pronounced fear of thunder usually spend energy and demonstrate considerable ingenuity so that they never have to confront the fear. This is a reasonable response, but it thwarts adaptive learning.

Trainers trying to get their dogs to lose their fear of something must ensure that the animals can't flee the scene. Treatments that expose the dog to the object or situation it fears in order to eliminate the fear are described as flooding, but this technique causes great distress and I don't recommend it. We'll look at more humane, but nonetheless effective, approaches in chapter 12.

Using unpleasant experiences in training

Instrumental conditioning based on negative reinforcement is when a dog shows an ability to bring an unpleasant experience to an end: for example, operating a switch that turns off a persistent loud noise. Negative reinforcement plays a big role in traditional heelwork training, where the unpleasant sensation comes to an end once the dog behaves (conventionally, walking alongside the handler's left leg). As with positive reinforcement, negative reinforcement must be immediate to be fully effective. For example, applying pressure to your dog's neck (via the collar and leash) to train it to walk to heel can work as long as the pressure ends the instant the dog complies. In skilled hands, pressure on the dog's neck can be so mild as to be almost benign and, with good timing, extremely transient. However, in less skilled hands, the delays and excesses of pressure mean that positive reinforcement is a better option for heelwork.

Similarly, using choke chains in dog training can only be humane if their characteristic sound acts as a warning to the dog. If the dog ignores the warning, pain will usually follow. Sadly, these dangerous devices are often not used properly (i.e., the tension on the leash must be released as soon as the dog responds appropriately by no longer pulling). Instead, some owners tend to hang on and launch into a miserable series of tug-of-war tournaments, frustrating for the humans and painful for the dogs. What's more, sometimes the choke chains are fitted incorrectly, which stops the chain releasing and prevents the dog's automatic relief from neck pain.

Learned food aversions

A dog develops a *conditioned taste aversion* when it tastes something—particularly something new—that causes nausea. This reaction evolved to help Uncle Wolf avoid dangerous but seemingly delicious foods. In the laboratory, nausea is normally induced by injection of lithium chloride, an otherwise harmless salt. Dogs can also be conditioned to avoid a place they associate with nausea, but taste-nausea associations are learned much more readily—often in a single trial. While most animals learn very readily to associate places, sounds or sights with shock, associations between tastes and shock are much weaker and variable.

Associations can form even though there might be a long interval between events. Many animals, including humans, learn to avoid foods that cause nausea even when the nausea takes hours to develop. Dogs are particularly good at learning to avoid foods that cause nausea, even if it arises some time after they've eaten, as long as the food responsible has a reasonably novel flavor. As opportunistic omnivores, dogs regularly encounter novel foods. No surprise then that, along with rats, they are world leaders in food-aversion learning.

CHEW ON THIS

A dramatic example of food-aversion learning comes from a study in which two wolves were given lamb meat laced with lithium chloride and wrapped in fresh sheepskin. The wolves acquired an aversion to the smell of sheep. Afterwards, when a live sheep was put in their pen, the wolves backed away as soon as they detected its scent. The sheep was soon chasing the wolves around the pen.

Does punishment work?

One note before moving on: To avoid confusion, learning theorists tend to use the term *avoidance* for training that requires a specific response to prevent an unpleasant experience and *punishment* for training in which only performing a specific response brings about a negative outcome. In everyday usage, punishment is usually delivered by one individual to another. Although it can be as mild as disappointment, it often takes the form of pain. Feral Cheryl uses it to prevent unwanted nipple nibbling by her growing pups. And legend has it that brave Uncle Wolf used it to send a usurper packing. In learning-theory terms it refers to anything that reduces the frequency of a response by punishing the behavior with a negative event. So, by definition, punishment always works. Whenever a response appears to drop out of a dog's repertoire, there is a chance that it has been punished. It is possible in the absence of a trainer to keep dogs in yards without real fences. Invisible fences can be used, whereby underground wires trigger a shock from the dog's collar whenever the dog starts to cross the boundary. So an attempt to cross the boundary is punished by a shock without the involvement of a person.

Punishment can be highly effective, but it can also easily go wrong. Like positively reinforced behavior, punished behavior may depend on the presence of a certain stimulus. The dog that has been punished for barking may no longer bark when its owner is nearby but may let rip when it is alone. A dog that has been punished for chasing cars may stop doing so in the presence of its trainer. An effective therapy program will recognize and meet the dog's need to chase, and encourage the dog to seek out its owner to help meet this need (such as an owner-centered ball game for dogs that chase joggers).

This helps to explain why it's so important in animal behavior therapy to identify what motivates unwelcome behaviors. Therapists can then suggest another, more appropriate, outlet for the animal to express its behavioral needs. Some punishment is clearly self-defeating, as when the punishment results in the exact response it's intended to prevent, such as hitting a dog that doesn't respond to a recall command, which will usually train it to keep its distance from the owner. More generally, punishing anxiety-related behavior (some kinds of barking) is likely to be counterproductive, since the punishment escalates the distress. And shouting at the dog for barking never works, first because it is not punishment and second because dogs could be forgiven for perceiving the shouting as the owner's best attempt at barking along with them.

Punishment can be extremely mild

Remember that in the current context, punishment simply means any event that makes a response less likely in the future. So it doesn't have to take the form of physical abuse. Even a verbal reprimand can make a behavior less likely, but it should be used with a command that offers the dog an alternative activity for which it can be praised. So, effectively the command "No!" (telling the dog that persisting in the current behavior will bring no rewards) is quickly followed by a command to do something else. There are more details on how to make "No!" an effective tool in your training toolkit and why the word need never be yelled in the next chapter.

Trainers of marine animals understand the effectiveness of strategic non-rewards. If, for example, a dolphin starts to mess up a training session, it may be sent back into its quarters (a punishment procedure involving a time-out, when the animal is removed from the training opportunity or just loses the trainer's attention.) This can be very effective as a mild punisher to reduce unwanted behavior. Clearly, time-out can't be effective if you first have to catch the animal before banishing it. If you chase a pup before expelling him, he learns that you are a hands-on trainer and that it's best to avoid you.

The problems with punishment

There is a pervasive problem with punishment procedures. Negative emotion due to the unpleasant experience—whether an electric shock, a loud noise, a sharp pain, or a blow from a stick—can become associated either with the immediate source of the punishment, human or otherwise, or with the particular place it happened. So various events compete for association with the emotion. This is evident in the *blocking effect* and in the *overshadowing* of one stimulus by another, as discussed in the previous chapter and in chapter 12 (Fine-Tuning). Which association ultimately wins out depends on a number of factors, including their relative novelty. A notorious example is the unfortunate association dogs make between livestock and the pain of electric shock collars used to punish dogs for chasing stock. This association can make dogs fearful of their former targets to the extent that they bite the cattle as a form of defense. Clearly, the risk of children inadvertently becoming targets of this sort of fear aggression is partly why few vets recommend the use of electric shock collars to novices.

Even under laboratory conditions it's not always clear whether punishment teaches the animal to associate a specific response with the unpleasant outcome. Instead it may simply cause general fearfulness that suppresses all behavior in the particular context. Encouraging the dog to give an alternative response makes it more likely that you reduce only the punished behavior, and it also makes it less likely that the unwanted behavior will return if the punishment is no longer used. Just like any other learned change in behavior, the discouraging effects of punishment can become less effective when the mechanism that brought about the change is no longer present.

Weakening the effectiveness of punishment

The effectiveness of punishment can be undermined in two more ways. The first is inconsistency—no surprise there! If the behavior is sometimes followed by a bad experience but sometimes not, it's likely to continue. The second, which causes the behavior to become very persistent, is when the punishment is initially mild but then grows in intensity as the behavior is repeated. For example, using a scolding cry that is quite soft at first but becomes sharper and stronger if the behavior continues, increases the need for more unpleasantness. Many of us have seen this in action with dogs that are reprimanded mildly at first and then need more and more emphatic, voluminous corrections that seem to be decreasingly effective. This occurs because some dogs welcome any attention, and, for all dogs, the shock value of the raised voice diminishes over time. I'm reminded of the typical Englishman abroad who, rather than using different words or, God forbid, the native language, invariably raises his voice to in an effort to make himself understood by non-English speakers.

Negative reinforcement and negative punishment

Good news and bad news are always relative on a sliding scale. So, of course, you can punish an animal by removing a reward. That's what we call negative punishment. And, of course, you can reward an animal by removing discomfort or pain. That's what we call negative reinforcement. The key here is to remember that positive or negative reinforcement always makes a response *more likely in future*. Positive or negative punishment always makes a response *less likely in future*.

Punishment and negative reinforcement are interrelated. The term negative reinforcement is almost politically incorrect, but in this context, negative refers to removing something from the animal's world, while positive refers to adding something. While many trainers claim not to use negative reinforcement, it's clear that when they reinforce a behavior by removing something unpleasant, they make the behavior more likely in the future; that is, the response has been negatively reinforced. By definition, if a stimulus is unpleasant, removing it is reinforcing. So assuming that pressure around the neck is always unpleasant, anyone who leads a dog with a collar round its neck is using negative reinforcement. Equally, in order to use negative reinforcement, a trainer must have used positive punishment (albeit mild) as well.

Negative punishment, or omission, helps improve or modify responses. During training, dogs usually offer an established response first. No reinforcement at that point (i.e., the trainer reserves the reward—a critical aspect of shaping) makes it unlikely that this now-unwanted response will be repeated. Reinforcement has been omitted (i.e., the dog has been negatively punished), which makes it more likely that it will respond in new ways. The trial-and-error process continues.

I mentioned the use of training discs in chapter 4. Plainly, their effectiveness depends on good timing. Like discs, the cue "No!" can work as a secondary negative punisher if it has gained its effectiveness by becoming associated with the removal of something positive—if only praise or attention—that the animal would normally expect. Outside this framework, yelling "No!" can be meaningless and entirely unproductive. Indeed, as Peta Clarke, a renowned Australian animal trainer, puts it: the more "No!"s you yell, the less you have trained your dog.

Physical punishment is neither effective nor justified, but punishment is not a dirty word, nor is negative. Both negative and positive punishment can be extremely mild. What is crucial is the degree and consistency you use to apply reinforcers and punishers.

Avoidance learning

For dogs, as for most other species, fleeing is the most common way to avoid unpleasant experiences. As we've seen, leashes are one of the most important impediments to a clean getaway, and their effect in developing a panic response is far too easily underestimated. One of the first questions you should ask when recalling a fear response in your own dog is, *Was he on the leash at the time?* If the answer is yes, then you know why you should expect to wait some time for the fear to subside.

Unfortunate laboratory dogs given shocks over which they have no control become emotionally disturbed. They learn that there is nothing they can do to switch off the unpleasantness and are said to have developed *learned helplessness*, a term used to describe the apathy that accompanies this passive coping mechanism. Perhaps this is why dogs that have had choke chains (here I go again!) fitted incorrectly—when the chain permanently acts as a ratchet rather than ever releasing to provide negative reinforcement—may become more and more difficult to handle and less enthusiastic about life. Dog owners who want to excel as life-coaches for their dogs should look out for signs of learned helplessness and stop whatever unpleasant experience triggered them.

CHOICE CUTS

- When it comes to punishment, it's easy to make mistakes. Take care.

- Dogs are very sensitive: A single bad experience can overwhelm previously conditioned responses.

- Dogs will learn to fear a stimulus if it reliably brings on an unpleasant experience.

- Once a dog has learned to fear something, it will show fear of similar things.

- Negative reinforcement occurs when the unpleasant sensation comes to an end once the dog behaves appropriately. This can be effective in heelwork training, but skill is needed.

- Choke chains can be frustrating for humans and painful for dogs.

- Dogs are world leaders in food-aversion learning.

- The effectiveness of punishment can sometimes be short-lived and be outweighed by negative consequences.

- It's really important to identify what motivates unwelcome behaviors. The response may be driven by an instinctive need—such as the need to chase.

🐾 Punishing anxiety-related behavior such as barking is unlikely to be effective, as it escalates the animal's distress.

🐾 The use of punishment often increases the need for further unpleasantness in the future.

🐾 Both negative and positive punishment can be extremely mild—and very effective.

🐾 Removing an unpleasant stimulus is reinforcing.

🐾 Exposing a dog to unpleasant events that it can't control can cause them to become unresponsive and hard to train.

Food can be used to help dogs overcome fear. Often called counter-conditioning, this strategy works best if the food is of high value and the fear is not overwhelming.

12
Fine-Tuning

Apart from suckling, chewing and swallowing, almost any motor pattern can be learned and therefore refined by training. The responses your dog ends up with are the ones you have reinforced or have allowed the dog to find reinforcing. So generally, when things go wrong you must blame yourself—the dog is never wrong. After all, he is behaving in a truly canine way and that, for him, is very appropriate behavior!

Obstacles to learning

Classical conditioning is selective. As one particular stimulus (say, the clatter of a metal bowl) becomes strongly associated with some important event (say, food, because the metal bowl is all the dog is ever fed from), the associations between other stimuli (say, the sound of a bowl formerly used) and food weaken. After Feral Cheryl decimated the local rabbit population, she had to start picking off new-born lambs, so the smell of rabbit became less relevant and was eclipsed by the whiff of tepid sheep afterbirth. And if your dog used to head for the front door whenever you picked up the car keys, he will quickly stop doing so when you get a 4WD for all those trips to the park and the beach. The new vehicle's signature sound rapidly becomes more relevant to him.

An important example in training is *overshadowing*. You may intend your dog to associate a certain word with an appropriate response, but if you accompany the spoken word with an unconscious hand gesture or body movement, the latter may instead become the effective trigger. In other words, body language signals have overshadowed the spoken signal or command. Consequently, they must either become the cue in your mind or be made less effective by making them less relevant or more variable.

Who's training whom?

When Professor Mills from Lincoln University, UK, asked dog owners to report any unwelcome behaviors and how they dealt with them, the responses he got (see the table below) raised many academic eyebrows.

Unwelcome behaviors reported by companion dog owners

Unwelcome behavior	% reported	% of dogs reported that stop when told	% of dogs reported that stop when given attention
Restless when they traveled	23.0	56.0	60.2
Lick or nibble themselves until sore	12.0	74.7	64.4
Chase themselves round in circles	14.5	71.4	68.6
Bark or howl incessantly	6.1	52.3	68.2

Two aspects of these responses leap out at me. First, 60 to 70 percent of dogs in this population have trained their owners to give them attention on demand. Second, barely more than 50 percent of dogs can be stopped verbally when either restless or barking incessantly. We have to question the success of the commands to stop and how well owners can actually train them. If these commands were truly effective as punishers, they would reduce the frequency of the behavior over time. This is probably why some animal behaviorists believe it's better to focus entirely on the power of the positive and say, "Never say no!" They don't accept that owners can be trusted with even the mildest punishers. I use "No!," albeit sparingly, because I can see it making unwelcome responses less likely in the future. For my dogs, it means stop what you are doing—it's not going to be rewarded. It is just another tool in my toolkit. Often, but not always, I accompany it by commands to perform alternative behaviors. When I am *free-shaping* a dog, letting it work out where the rewards are to be found within a particular challenge, I gently utter "No!" instructively to divert him from following his current line of enquiry and shuffle him along to try something else.

Another aspect of these responses that surprises me is the low number of barking dogs. Barking is universally the most common cause of complaints to local councils (not uncollected litter bins, not badly maintained roads, not even exorbitant rates). The fact of the matter is that most owners of barking dogs unwittingly underestimate the gravity of the problem because, of course, they are absent when most of the barking occurs. If the neighbors had been included in the Lincoln study, the percentage of incessant barkers reported may have been higher. Either way, getting dogs tired before they are left alone is a good way to reduce unwanted yodeling. One of the most innovative ways to achieve this is to check out finding a back-yard buddy for your dog.

Removing rewards

Classically trained responses become progressively weaker if, repeatedly, the conditioned stimulus is no longer followed by the unconditioned stimulus. We saw this with the above example of the new 4WD dogmobile: The change of car and change in the sound made by the new set of car keys quickly became more relevant to the dog. Similarly, extinction occurs when a response learned via instrumental conditioning is no longer followed by any reward. The response will be performed progressively less frequently and less energetically. A dog that begs at the dinner table will stop begging if never rewarded. (Believe me!) However, during extinction of instrumental behavior, dogs often revert to innate or previously learned behaviors. So be strong! The same principle allows you to tidy up the rough edges of a learned response, such as a sloppy sit or a complacent come. If you withhold the reward, your dog will experiment with alternatives, and your job as a coach is then to shape improvements in the right direction.

There can be some intriguing outcomes when a previously rewarded response becomes extinct: Things sometimes get worse before they get better. So, again, be strong! Early in extinction programs, a *frustration effect* is common, so for a short time the response is made more vigorously than when reinforced. A pup that has learned to slip its collar to get free will struggle much harder when first tethered to a properly fitted collar. It's important to be aware of this effect so you don't leap at early conclusions that your program is a failure.

That said, a much more effective way of eliminating an unwelcome behavior is to reinforce all other behaviors that arise in similar situations. For example, if I want to stop a dog jumping up at visitors, I may reward it for every, single non-jumping response it makes around unfamiliar humans. This is called *omission training*:

Rewards are withheld if the target response is made, and thus they become contingent on the target response *not* being made. So the rewards stop if a jump is offered. Such an omission schedule can be used to change unwelcome behaviors, and it is preferable to withdrawing reinforcement completely, which would run the risk of removing all incentive to respond.

Just as it is important to avoid confusion and promote creativity when training a new behavior, it is imperative that, when training a dog to stop performing some problem behavior, you simultaneously give it the opportunity to perform a more acceptable behavior with a similar motivation or a contrary body posture. As described in chapter 11, a dog that chases joggers can most easily be trained to stop and look at the handler if it associates the sight of a jogger with an owner-centered ball game. Meanwhile, a dog that barks incessantly can be showered with rewards for lying down (a postural response that happens to make barking difficult) and not barking.

During omission training or extinction, all context-specific stimuli can exert considerable control over behavior. As I have assured you already, the dog that begs at the dinner table and scavenges food from poorly trained or mischievous children will stop if no further tidbits arrive. If grandmother is the one who stops the flow of tidbits, then extinction will occur most quickly when Granny comes for a visit. Begging may return, at least temporarily, after she leaves. The message here is that consolidated extinction training must be conducted in numerous contexts.

Partial reinforcement

So far my examples of instrumental conditioning using positive reinforcement have involved rewards being delivered after each response, that is, *a continuous reinforcement* schedule. A response will be learned faster this way than with *partial reinforcement*, when only some correct responses are rewarded. On the other hand, partial reinforcement can make it harder to eliminate unwelcome behaviors. Dogs that are *sometimes* given tidbits for begging take longer to give up when owners learn never to reward the behavior than those that have had constant reinforcement.

Trained behaviors can persist even when the benefits of performing them no longer outweigh the costs. For example, Skinner trained an extreme example of a rat on a reinforcement schedule so stingy that the energy received from the food was less than the energy expended in getting the food. If the balance is right, animals will continue to perform behaviors even if doing so causes them some discomfort. The same principle is illustrated by canine members of dancing-with-dogs (also known as canine freestyle or heelwork-to-music) partnerships. They exert energy offering responses that were once rewarded frequently with food but have subsequently been linked together. The responses cannot be reinforced individually because of the need for continuity in the dance as a whole. In fact, for most of these dogs, the pleasure of leaping around with their handlers while enjoying their undivided attention soon eclipses the value of all the morsels used to train each step. The sum of the whole eclipses the sum of the parts that built it.

A strong recall response can save your dog's life

Having said that, when you really need a reliable response every single time, you have to ensure that the dog's compliance is not based on whether he feels like being good. An excellent example is the recall. Coming when called can save your dog's life and, if he is racing off after some wildlife, the lives of other animals. Every single owner who brought a mashed-by-car dog to me, while I was in general practice, seemed surprised that their dog had been run over; most of them reported that the dog had never before run into traffic while being called. So, my single piece of advice here is: Don't let there be a first time. Your dog's recall has to be a nonnegotiable response, because when dogs are heading towards traffic, nanoseconds count. This is the single most important reason why you should never utter your dog's name angrily, no matter how wound up you are. He has to know that the sound of his name is good news, his life-coach heralding yet another great opportunity.

Some dogs in a park seem to associate being called with the end of the walk, and, not surprisingly, they suddenly acquire temporary deafness. Others work out that being called to their owner's side is associated with the presence of delicious distractions. If your dog looks around him when you call him during a walk and seems to be trying to see whatever it is you are determined he should not chase, then his recall definitely needs fine-tuning. If these unwelcome responses sound familiar and remind you of your own dog, you should probably consider changing the recall command completely (e.g., from "Come!" to "What's this?") because the current command is rather unreliable. You'll be delighted with the change if you take a pocketful of treats, including some jackpots, with you from now on.

SEVEN TIPS TO IMPROVE A RECALL RESPONSE

- Reward continuously until your pup is (skeletally) mature.

- Reward intermittently (partially) after that age until his dying day.

- If he doesn't seem pleased to return to you, increase the frequency and quality of rewards (the reward ratio) immediately.

- Include some amazing, delicious, best-ever jackpots that are reserved for recalls.

- As your dog gets close, run back a few steps. This trains him to get right up to you. In a crisis, this is important because it allows you to grab him easily, if necessary.

- Use a release command, such as "All gone!" or "Okay!," so that your dog does not create his own cues to return to what had interested him.

- Use this cue to allow the dog to return immediately to some items that have grabbed his attention, otherwise he will regard returning to you as just plain dull.

The importance of context

Pavlov's dogs may have had their movements closely restrained for experiments, but they knew that the lab was where they received meat powder. The straps that held them in place on the bench-tops may well have caused great resentment in another context, such as a park. However, their effect was context-specific: While white coats represent danger to some dogs at a veterinary clinic, white-coated visitors to the home don't provoke a similar panic response. Puppies who behave impeccably at training school but apparently forget everything when out on a walk also show that learned behavior can depend on a specific context. Good coaches reduce this sort of context specificity by varying the settings in which they train the puppy. For example, one of the most time-consuming elements of guide dog training, after basic training with artificial obstacles, is the pro-cess of repetition in various contexts to eliminate dependence on the training ground's environmental cues.

Discrimination training

When a dog performs one response after a specific command and only after that command, the response is said to be under stimulus control. In some kinds of dog training (e.g., for movies), it doesn't matter if the response is triggered by signals that are similar to the trained command. Indeed, this may sometimes be an advantage. For example, a trainer may want a verbal command to be effective when other people (such as actors) use it, even though the actual sound may be very different. However, generally speaking, more precise stimulus control is usually required. Remember, most training outcomes depend on consistency.

It is usually best to train a discrimination using positive reinforcement. In *simultaneous discrimination* training the dog is given access to two or more stimuli at the same time and is rewarded if it responds only to the target, or positive, stimulus. For example, if you want to train your dog to discriminate a blue ball from balls of different colors, you might start by rewarding it for interacting with the blue ball alone and then for approaching when it is placed beside a white ball. Within a relatively short time the dog will pick up only the blue ball (the instrumental response has come under control of this color). Establishing precise control over picking up the blue ball can be achieved by providing an array of three balls—one blue, one green and one blue-green—and rewarding only for picking up the blue ball. As in all forms of simultaneous discrimination training, it is important that the position of the blue ball varies very frequently, appearing on the left, right and in the center. Otherwise the dog will almost always learn *where* to respond to rather than *what* to respond to (it tends to develop a position preference that overshadows learning the visual cue). You will have noted that the dog in this example first had to discriminate between a blue ball and a white ball. Only once this was established did the discrimination task become more complex. In general, such *easy-to-hard procedures* make possible far higher levels of performance relatively quickly on a difficult discrimination.

Sniffer dogs are trained in this way, by rewarding responses to correct stimuli and not responses to all other stimuli. However, if the discrimination is a difficult one, the dog may continue indefinitely to identify the wrong stimuli, even if at a lower rate than responding to the correct stimulus. There are many real-life situations in which a dog's discrimination performance has to be near-perfect. For example, airport managers could not cope with criticism because innocent passengers were being targeted too often by dogs trained to sniff out narcotics or explosives. One common system of training for this outcome is to reward the dogs for not making the target response to negative stimuli. This is an example of an *omission schedule* of reinforcement, as discussed earlier, in the case of dogs being rewarded for lying down instead of barking incessantly.

Verbal commands and other signals

Many trainers use specific noises to trigger a certain response. An auditory trigger can be any sound, as long as it's distinct. Issuing clear signals is crucial in dog training, given the importance of consistency. Use critical signals only in particular circumstances. A good example is the release command used by handlers of police siege dogs when they send their dogs to attack.

By training animals to respond reliably to signals and then making the signals more and more subtle, a trainer can elicit behaviors without an audience perceiving how. Once the animal responds to only the tiniest cue, the trainer can introduce a spurious "command," given without subtlety, so that the animal appears to understand the nuances of human language. Poachers and their dogs traditionally used a similar approach to confuse their pursuers. Since to seize the poacher's dog was to identify the poacher, gamekeepers who stumbled across a shifty looking man-and-dog team were keen to catch both players at the scene. Often the human vanished before the dog, and so the gamekeepers had to try to persuade the dog to approach them. This was why poachers trained their dogs to run away whenever coaxed to approach in the usual way, with commands such as "Come here, good boy!" By using novel words to recall his dog, the poacher could remain slightly ahead of the game, as it were. A related off-switch appears in sheepdog training. When a shepherd works a brace of sheepdogs, he must teach dog A that any cue prefixed by dog B's name is a command to keep performing A's current behavior.

Being consistent when issuing commands is critical because it is the very best way of not confusing your dog. So, once you have decided on a command for a certain behavior, stick with it. Changing a command is usually advisable only if the dog has been retrained to offer a different or better response. A common mistake is to issue a series of commands that change, sometimes slightly, sometimes radically, from one to the next simply because the first didn't work. This is an approach humans use when making themselves clear to one another, but something that may seem an obvious connection between one command and another to a human is never as straightforward to your dog.

The key here is to be absolutely certain about what your dog does when it hears a given command. I say this because so often in veterinary practice I see owners using a series of multiple commands for a single response. For example, they come into the clinic and want to show me that their dog can sit on the floor, so they issue a string of commands, one after the other (often without even a pause to see how the dog responds):

Jasper, sit!
Sit, Jasper!
Jasper, sit down!
Good boy, sit!

All of these sound very different to poor wee Jasper, who usually gets his collar yanked up or his bum pushed down (two further commands but this time from the hands-on school of training techniques). He inevitably learns to ignore the owner's meaningless drivel. Consistency has gone right out the window.

It is easy to see why families with small children can rarely give consistent commands and so often struggle to train their dogs effectively. The tendency is for uninitiated members of such families to issue commands differently, using various intonations, prefixes and even different words. In addition, food is readily available as toddlers toddle around carrying biscuits and chips. While some dogs hover like rock spiders with prey in their sights, others simply learn to "accidentally" knock food out of small hands and gobble it up almost before it hits the deck. So dogs quickly learn that crime, in this instance mugging children for edible booty, really does pay.

INTRODUCING A COMMAND

One way to fix a command in place is to use it for a new response throughout the training of that response. For example, it seems obvious that to train your dog to stay for 10 minutes with you out of sight, you must first train it to stay for 30 seconds while you are still present. So, the dog hears the command before it offers early attempts at the desired response. However, there is an important drawback with this approach, even though at first glance it seems to epitomize consistency. Unfortunately, because the dog gets rewarded for good attempts, it may learn to link the command with less-than-perfect responses. This can increase the time taken to train the perfect response because eventually the dog has to unlearn the less-polished response to the command.

A further example comes from training a pup to sit. If, when told to sit, he lowers his hindquarters and crouches (or half-sits) and is quite rightly rewarded for the effort in the right direction, he may continue to offer half-sits instead of immediately sitting properly when he hears the sit command. So, a specific cue used early in training can train the dog to expect a reward for substandard approximations of the desired behavior. The dog must then learn to work harder for the final shaped behavior by a process of extinction (learning that poorer responses are no longer rewarded).

As we mentioned in the section on clicker training (page 176), a practical solution to this potential problem is to train a new response until it is perfect and your dog is offering it all the time, and only *then* put it under stimulus control by adding the chosen command. It seems that a verbal command can be included at any point in the training process, but there is a good argument for using a specific command only when the final behavior is being offered. Doing so allows you to establish a distinct trigger for the *polished* behavior rather than training your dog to respond to the cue with a half-baked response. With the desired behavior under stimulus control, the command prompts the perfect response every time.

Here is an example. When coaching a dog to look away from me, I might shape him to turn his head to his right and I might move my hand in that direction to prompt him. Once he starts offering a small but convincing head turn with every prompt, I reserve the reward at each step until the turn becomes more marked and the response more immediate. This is the easy bit because the dog offers the same response spontaneously, time and time again, as he is confident that I will pay him with rewards. Once the head turning is rapid and emphatic and I feel it has reached my training criteria, I give him the cue "Not looking!" and stop using the hand signal. Then I reward the dog *only* when I have given the "Not looking!" command. Spontaneous head turning is not rewarded. Head turning is now under stimulus control, and whenever the dog is drooling in front of a house guest as she eases a delicious biscuit towards her mouth, I can say "Not looking!" and the dog looks away as if consumed with shame.

Trained dogs learn to discriminate not only between general conversation and commands, but also between one command and the next. Although it would be absurd to isolate companion dogs to facilitate their training, professional trainers often limit the time they spend with working and obedience dogs. Isolating animals from trainers when trained responses are not required can be effective in reducing *generalization* (animals giving trained responses to cues similar to those used in training). For example, if a championship obedience dog hears, say, the last syllable of bis*cuit*—perhaps during a teatime conversation—and responds with an immediate and impeccable *sit* response that goes unpraised, he might eventually stop offering the trained behavior. Sheepdogs that spend most of their days chained up or in crates are very keen to work when emerging from confinement. They can be certain that any signals are meant for them.

In advanced dog-obedience competitions, using the voice attracts penalties in demonstrations of control at a distance, so trainers use hand signals that have been paired with spoken commands earlier in the dog's training. Visual signals are often used from the outset of training. A common example is in training dogs to sit when they see the owner's hand being raised. There is some merit in this since it can fortify the trigger to sit on a windy day when the dog is working at a distance or even as the dog ages and becomes deaf. Having said that, the use of two commands of a different nature during the early formation (shaping) of a response runs the risk of overshadowing. One response for one signal makes training easier for both you and your dog.

COACHING A SEQUENCE OF BEHAVIORS

By linking trained responses (*chaining*), a dog can be coached to perform a long sequence of behaviors where there may be some time between the initial response and the reward that follows the last link. One of my own dog's tricks is a good example of chaining: Wally waits while coins are thrown around, then advances to each of them, stacks them in his mouth, drops them in a bucket, returns to pick up any coins he could not retrieve in the first mouthful, and only then picks up the bucket by its handle and brings it to me. Each of those tasks was trained separately before being chained together.

Chaining can be very important for traveling along complicated routes where landmarks—visual cues and smells—can serve as triggers for each part of the journey. The most important point about any kind of training a chain of behaviors is to start with the last step and build the chain backwards. For example, in training an animal to negotiate a complex maze, the trick is to begin training at the last choice point. Wartime messenger dogs were trained to race back to their handlers over gradually increased distances. The reward, a piece of liver from the trainer at home base, was always the same, but in each training session the first part of the mission was novel and challenging. Once on their familiar home track everything was plain sailing.

Ending on a high note

Knowing when to call it a day is a critical skill in dog training. Training manuals stress the importance of always ending a training session on a high note—but why? Imagine Uncle Wolf trying his best to catch that rabbit. If he tried a new, improved, extra-smart move and it didn't land him his rabbit-flavored jackpot, he would not bother with it ever again. You don't want to lose the brilliance in your dog's performance by failing to reward it or griping about how it wasn't quite good enough. Always stop well before the dog gets tired and his enthusiasm starts to wane. If you continue past this point, you may have to reward the dog at the end of the session for a poor performance. From this, he may learn that the premium versions of the behavior he offered earlier in the session were not necessarily required.

CHOICE CUTS

- Be careful of "overshadowing" your dog's trained responses with unintended signals that will only cause confusion.

- During a program of extinction of a previously rewarded response, matters may get worse before they get better.

- Omission training eliminates unwanted responses by rewarding all other behaviors in the situation where the unwanted response tends to occur most strongly.

- Dogs that are sometimes given tidbits for begging take longer to give up the practice than those that have had consistent reinforcement.

- Good coaches vary the settings in which they train.

- Once a response appears immediately after the stimulus and never after any other stimulus, it is said to be under stimulus control.

- Clarity of signals is one of the most important aspects of good coaching.

- Train one response for one signal.

- Build a chain of behaviors by establishing the last step and developing the chain backwards.

Feeding a sniffer dog as close as possible to the spot where it responded correctly to a target odor helps to fortify classical associations between stimulus (in this case, odor) and reward.

13
The School of Life

Learning is essentially a form of problem-solving; it's what goes on when a dog has to find a way to get further into its comfort zone. When rewards are on offer, the problems are to do with *How do I get what I want?* and *How do I get it as soon as possible?* Where pressure is being applied (e.g., on the collar via the leash), they are to do with *How do I get rid of the pressure?* or *How do I make this end as swiftly as possible?* The School of Life teaches how to solve problems with an ever-decreasing number of errors.

Social skills

Learning to cope with changes in the social network is part of life in any gregarious species. Uncle Wolf had to adapt to social flux as strangers occasionally drifted into his pack, older members of his pack died off, new alpha males and alpha females stepped up to the mark and litters emerged from their birth dens. Life-long learning is as appropriate in the world of the companion dog as it is topical in the corporate sector of modern human existence. A dog's social skills may well be cast in the early months of puppyhood, but, in the best cases, they continue to be polished throughout life. Learning is all about adaptability, and dogs, those consummate opportunists, are supremely adaptable. They have to be because they can't afford to get set in their ways. Imagine Feral Cheryl scavenging along a village street day after day. If she didn't change her routine, the chances are she would starve because she'd deplete certain resources and never discover new ones.

A dog's value system changes with circumstances but also with time. For a young pup, play may be the most valued resource in life, but give it a year with its testicles intact and you'll probably find that sex becomes more important. A dog's needs change as it matures, so adaptability is essential in the transition from juvenile to adult. Indeed, even in their dotage, a dog must adapt to its diminishing social status. Combat becomes too difficult, even timing can lose its edge—a well-aimed lip curl may be entirely useless if its timing is all wrong, and ultimately deference becomes the better part of valor. Dogs that have been passive leaders seem to find this transition easier than those that have had to fight to maintain high rank. By being so very distinguished, they set inspirational guidelines for humans who strive to be benign group leaders and effective life-coaches. Their dignified passage through life offers a template of how to behave around dogs, but for many of us such deportment is far from innate. When our natural response is to react, defend and retort, we should probably relax, defer and reflect.

Puppy preschools

It is important to allow puppies time to be puppies. They take a while to develop the behaviors we expect of adult dogs. Having the comfort of other pups around is important, so we can see how the den is regarded as not merely shelter or where one is most likely to find food but also as the place where the other pups are. Excursions from the den are an early form of dispersal, but studies of puppies reared under semi-feral conditions have shown that until around 12 weeks of age, exploration beyond the den is limited.

Puppies generally play well together as they discover the greeting strategies of other dogs and, as a consequence, their own social strengths. Indeed, a few boisterous bouts seem capable of inflating some pups' confidence to the point of bullishness. Learned playground bullying is one unfortunate outcome of the type of puppy preschool with outdated protocols to facilitate puppy play. If we can agree that the best training is structured fun, the best preschools are those that place considerable emphasis on the structure rather than simply maximizing the fun. It is now recognized that when play takes the form of all-in wrestling sessions, it is bad news for many young dogs that lack the experience, body mass or inclination ever to end up on top. Thankfully, the days of pile-ups involving madding crowds of pupils in puppy classes are very much numbered.

There are scientific reasons why certain behaviors should be taught at certain times in a dog's life. Building confidence in pups should be a priority for the organizers of puppy preschools, along with educating owners to recognize the importance of consistency and the critical value of good timing. Since young dogs often lunge at one another, owners must be trained to take responsibility for their dog's actions and respond appropriately when their dog causes distress to classmates.

While puppy preschools undoubtedly offer a unique opportunity to teach new owners about dog care and health, their place in the early education of pups has been questioned by those who feel that pups are more likely to learn complex responses when trained either in isolation or in the presence of an older trained dog. That said, however, training pups to pay attention to their owners and handlers when surrounded by other youngsters may be enough justification for attending a puppy preschool.

One-on-one

Socializing one-on-one with your pup will allow you to deliver lessons with consistency and therefore greatest effect, even though it may not appear to be as much fun for your pup as free-for-all play with multiple partners. That said, puppies' capacity for fun—between power naps—is phenomenal and can rarely be sated by one adult, however playful and tolerant. Dogs have evolved to be surrounded by playmates. Adults and juveniles often take turns playing with pups, perhaps because they lack the energy to keep up and so regularly work as tag teams. This is appropriate, since it seems clear that pups are better able to learn the art of play—something that sets them up for life as proficient social athletes—if they have only one tutor at a time. Having a variety of "dog fathers" and "dog mothers" (not to mention surrogate or real siblings) will always be better for the pupils than having just one role model. Different games emerge with different play partners, so a richer spread of tactics will tend to emerge when a pup has several tutors.

Having a variety of playmates

Play with dogs of various age groups is also better than with just one, since the pup can perfect certain skills with each. For example, a game such as tug-of-war can be played more easily with a willing juvenile than with a serious adult who barks as soon as the pup approaches the article of interest. Meanwhile, agility and safety may be best learned from a seasoned expert member of the adult community rather than a gangly, naïve juvenile.

It is clear that dogs need fairly constant exposure to new doggy faces to maintain social decorum. If they remain in one patch for too long, they readily become territorial. The patch in question can be any area that offers useful resources so can include yards, gardens and, of course, parks. When dealing with dog–dog aggression, many behaviorists advise walking the dog in various locations to avoid it becoming territorial about a particular place. The strategy is to break down affiliations with a single park since it may become established as a territory that must be defended. This makes dogs more deferent with other dogs (as if they assume that they are guests in rather than owners of the new area).

Formal training

Formal training can be hard work, requiring time and mental energy. To make the most of it, be clear about your goals. These can be broken down within each training session into achievable sub-tasks. The sub-tasks translate into criteria for your dog to reach (see chapter 10 on shaping behavior and reaching criteria).

Mistakes made during formal training sessions can be frustrating, but those made in public can be embarrassing. I am reminded of the time when, as a schoolboy, my dog and I were competing in an agility competition. My beloved blue merle collie, Ben, stopped just before the see-saw to cock his leg against the steward's nylon slacks. I was mortified. For many dog owners, peer pressure is the worst aspect of dog competitions and the reason so many choose to avoid them. However, the emergence of training clubs has encouraged owners who might otherwise have stayed at home and trained their dogs in isolation. The presence of other dogs while training is pivotal here. With all dogs under control, each of them can learn that rough-and-tumble is off the menu for the time being. This certainly helps the dog and handler to focus. Owners can concentrate on shaping new responses and refining established ones, while the dog can be assured that all incoming signals are relevant to him.

Training your dog *in the company of other dogs that are under control* is the most valuable aspect of any dog-training club. Although you can't expect to be able to do this in a dog park, you should still be able to shape and reward the behaviors you like. There are so many clicker trainers in dogland these days that one click can attract a mob of expectant snack grabbers, which, in turn, can sharpen your own dog's attention.

Practicing what you have learned at a training club is the next key step. And being true to yourself is another critical ingredient in a successful recipe. There is no point kidding yourself that all the refinement will magically appear on the day of a trial. An ongoing video record is an excellent means of avoiding self-deception. Your practice sessions should show continuing improvement, and if they don't, you need to explore the reasons for stagnation or decline in the quality of your team's performance. Staleness in obedience dogs usually mirrors a lack of enthusiasm in the handler or inappropriate attention to (punishment of) minor details that make the dog reluctant to remain involved. It is important to know how to apportion blame to the variables in the training equation—that is, when to blame the dog (almost never) or circumstances (rarely) or oneself (almost always).

Especially at the outset of training, it is essential to maintain the fun. Don't train a dog to come and sit in front of you; rather, just train it to come and expect something great, such as a ball to be chased. Again, becoming a fellow adventurer in voyages of discovery gets far better results than being a formal trainer.

Learning with others

Humans are rather unusual in their efforts to teach—or train—children and animals. Indeed, we have been dubbed *Homo docens*—the teaching ape—and we may be the only species of which individuals deliberately instruct others. But humans are not the only species who learn by observation. Animals from some species learn a great deal from one another by a process known as social learning, which cuts the high costs of trial-and-error learning. Fear, for example, can be highly contagious. Monkeys can develop a fear of snakes simply by watching a video of another monkey reacting fearfully to a snake. Birds that see a member of their species reacting fearfully to a novel object will learn to mob that object even though it hasn't directly threatened them.

Learned food selection provides a good example of social learning. In the United Kingdom, in the 1930s, milk delivered to the doorstep started to arrive in bottles with shiny aluminum tops. One small bird, the blue tit, quickly learned to peck through these tops to reach the milk. Once a few individuals had learned this trick independently, the practice rapidly spread (via a process known as social transmission) to almost the entire population of UK blue tits. However, it is not clear whether this was a case of social learning (tits learning by watching expert birds perform). An alternative possibility is that learning occurred less directly, as a result of one individual setting up an occasion that made it easier for the next to learn, for example, by some birds reaching the cream through holes punched by others.

Young pups can learn that certain smells are important to their mothers, a finding that is now being used to accelerate the training of sniffer dogs. In a system developed by the South African police, trained bitches are exposed to the target scents in the presence of their litters. They respond by sitting attentively, and this increases the significance of these materials to the youngsters when they undergo formal training. Similar studies have shown that naïve dogs can acquire food preference by sniffing others. And recently, animal shelter staff, recognizing the appeal of at least some obedience in their adoptees, have demonstrated that training their inmates to sit on command is easier in the presence of trained canine demonstrators. Furthermore, there is evidence that dogs housed alone are cognitively impaired, although (as we have seen in chapter 3) this may simply reflect a general lack of optimism that emerges in unenriched environments.

In cases where information hasn't been transmitted socially, learning is based on first-hand experience. Expecting a dog to acquire some behavior simply as a result of observing members of its own species, let alone by observing human performance, will almost always lead to disappointment. When novice foxhounds are coupled (chained) with older members of a pack, they learn not so much by observation as by having to respond appropriately to avoid discomfort. The command to which the older hound responds comes to predict for the youngster some neck discomfort that can be overcome only by traveling in the appropriate direction. Eventually, the command works directly by eliciting the desired response immediately, and so the older hound, having done its job, can be released from the bonds that tie it to its student, an outcome that presumably brings considerable relief to both dogs.

Learning by copying

It is important not to confuse social learning with the less complex phenomenon of social facilitation. This is the effect one animal can have on others of its species simply by interacting with a resource, for example, when the sight of a chicken pecking at grains on the ground prompts observing chickens to do the same, even though they may not be especially hungry. Social facilitation, mentioned in chapter 3 (What Dogs Value) is when one dog observes a second performing a certain behavior and then copies that behavior. It explains the tendency to join in an activity (e.g., feeding, playing, barking).

One of the most familiar canine examples of social facilitation is grass-eating, which seems to spread like wildfire among dogs, especially when on a walk. One dog (the tutor) stops, sniffs (of course) and then starts teasing strands of grass into its mouth. The next dog (the observer) catches up and, almost without sniffing, starts grazing voraciously; the two apparently compete for the best blades. But what is it that defines the most desirable herbage? This question runs to the core of one of the most brightly burning topics of debate among dog walkers: Why do dogs eat grass in the first place? Most of us see a link between an episode of grazing and subsequent vomiting (often with something indigestible, such as mashed-up bits of a frisbee, in the vomit) and assume that dogs eat grass as an emetic. However, alternative theories for grass-eating abound: to supplement themselves with the magnesium found in chlorophyll; to add fiber to their diets; to expel worms from the lining of their guts.

CHEW ON THIS

A good friend of mine, Samie Bjone, has recently completed a PhD on grass-eating by dogs. At first glance this may seem an obscure topic, but when one appreciates that almost all dogs eat grass and no experiments have been conducted on it, the merit of her endeavor becomes clear. Among many other things, she examined the effects of gut disorders on dogs' grass intake to establish whether they are self-medicating. Samie's experiments demonstrated that dogs do not eat grass to self-medicate a naturally harbored worm or an induced mild gastrointestinal disturbance. She observed the social facilitation of grass-eating in pups as they picked up the behavior from their mothers. She also asked owners about the characteristics of grass-eating dogs in a bid to identify the so-called risk-factors for grass-eating. In other words, by working out which animals are most likely to eat grass, she hoped to be able to identify what motivates them to do so. All dogs in her study were in excellent health and all readily ate grass. So she deduced that grass-eating is both normal and common and concluded that it is neither a problematic behavior nor an indicator of illness. The right grass is really just an occasional snack. Dogs influence each other by zeroing in on the "right" grass.

Mimicking humans

Unlike other species, dogs sometimes seem to regard us as members of their social group and look where other dogs and other humans are looking (this is known as social referencing). It is interesting to speculate which human behaviors prompt attempts at mimicry from canine observers. Many owners believe digging in the garden is something that can be quite contagious among juvenile observers— and why not? Dogs love to dig, and they are social animals, so why not dig as a group? In reality, the actions may be similar, but the intentions of the two species are very different. Dogs can't be expected to understand the principle of putting plants in the garden when digging bones out of it would be far more appealing.

Drinking is another contagious behavior among dogs, most probably because, for free-ranging dogs, water is not necessarily always available, and so once it's been found, it must be accessed. Once one dog starts to drink others often follow, but humans drink so differently from dogs that our behavior is not readily mimicked. Dogs see us put vessels to our mouths but can't be sure we're drinking, since we do so without any of the characteristic slurping they might associate with quenching thirst. Eating, on the other hand, is socially facilitated in humans and, most probably, in dogs. Begging at the table is a simple canine response to seeing the pack eating (although it is quickly reinforced when tasty snacks appear table-side).

Dogs are outstanding students of human behavior and are adept at learning from humans socially. They can even acquire a new habit without extrinsic reward or social feedback. For example, they can learn to take an efficient detour by observing a human doing so. They have been shown to learn new ways of manipulating objects (such as pulling a bar with a paw instead of the mouth) after watching human demonstrators and to match their behavior with human action sequences. Individual dogs' performances in so-called "Do as I do!" tasks are comparable to those of great apes. These findings show that some human behavior can be a model from which dogs can learn. Unfortunately, our skill in training dogs through social facilitation is limited. This is where our ability to act as leaders of dogs is most lacking. Assuming we are good at gaining a dog's attention (see the section "Do I have your attention?" on page 119), we can encourage him to walk by walking, run by running and lie down by lying down. But we are kidding ourselves if we consider that sitting still for 10 minutes is going to train any observing dog to perform an extended sit-stay. When we encourage interaction with a ball, by playing with it ourselves, we are primarily adding value to it as a possession—a form of stimulus enhancement (see chapter 10, on learning about opportunities). I can offer a similar cross-species example with one of my own dogs: Wally was watching me target-train Hector, one of my horses. The target was a section of white tape at the end of a 1-yard-long black plastic pole. Always keen to jump the line for any available resources, and without any experimentation or prior experience of this apparatus, Wally targeted it repeatedly with his nose as he had seen Hector do. The target's relevance had been enhanced by Hector's use of it. And yes, I rewarded Wally for his resourcefulness!

Do dogs know what another dog is thinking?

If social facilitation brings advantages to those animals that follow the examples, perhaps some leaders see the advantages of *avoiding* sharing resources, in other words, for them to use social facilitation to mislead observers. Deception in dogs implies that one dog can work out what another may be thinking. It relies on them having a *theory of mind* that attributes knowledge to certain individuals and makes them therefore worth observing.

To test whether an animal has a theory of mind, we need to set up an experiment that allows one dog to witness events that some dogs can observe and others cannot. The principle is that those dogs that have watched the events will behave differently from those that have not. Crucially, dogs that see the events going on are more likely to use the informed dogs as sources of information. Usually these studies and the information exchanged within them revolve around food. The (test) dog being studied to see whether it has a theory of mind is allowed to learn that a human repeatedly hides a food item in full view of one (informed) dog and partial or obscured view of another (control) dog. If the test dog has a theory of mind, it will use the informed dog's behavior to modify its own hunt for food later in the experimental environment. So far at least, tests have failed to confirm this ability in dogs, while pigs seem to have it.

CHEW ON THIS

Both insight and deception are of tremendous interest to an emerging group of scientists who often call themselves cognitive ethologists. These researchers look for evidence of higher mental processes in animals. For example, they study how observation affects the behavior of observers, noting the relationship between the demonstrators and the observers. Chickens, for example, can be influenced in their selection of food by observing the choice made by a leader as opposed to a regular peer. So the choice of demonstrators in dog versions of these studies may prove fascinating.

Insight is when problems are solved without trial and error, while true deception depends on the deceiver having some notion of its victim's mind. So, can a dog deceive another in making a choice that gives it an advantage? An example is when one dog's access to food is being made difficult by a second dog. The first dog runs toward the door barking, and the second dog runs to join him, leaving the food unattended. The first dog then returns to the food at speed. Many owners report this finding with amusement and some pride. The apparent guile is impressive, but remember that this response could, of course, have been learned by accident or, more correctly, through trial and error. But if it occurred spontaneously without any shaping, then it's good evidence that dogs have insight. The trouble is that these anecdotal accounts are notoriously hard to verify. Despite the difficulties in finding compelling evidence of a theory of mind in dogs, it's a noble quest. After all, if one dog can know what another dog is thinking, it may have a sense for what it is feeling, and this is the critical step for it to have the quality humans prize as compassion.

Choosing a command

Balls, bowls, leashes and car keys are all very familiar triggers for behaviors that are learned with little effort. As opportunists, dogs are constantly looking for clues that make life easier and more fun. In dog training, commands and cues are the labels we use for triggers. Essentially, they are telling the dog "here is an opportunity—if you perform behavior X now, you will be rewarded." The command helps the dogs identify or discriminate the opportunity, and that is why triggers are more formally called discriminative stimuli (see chapter 12).

Any signal can be established as a trigger. At the risk of stating the obvious, it's important to select triggers (discriminative stimuli) that are easily discriminated, not just from one another but from general conversation. Voice commands are the most common triggers, and you can choose any words or sounds to signal for behavior X. Having said that, it's helpful if they are not too similar to other commands the dog already knows, since they can prompt these behaviors by mistake. The importance of keeping one command for one response, and consistency in general, was emphasized by the example of poor wee Jasper in the previous chapter. It also helps to avoid sounding gruff or stern. Keep commands upbeat, and you'll find them easier to say with a playful tone of voice. Dogs thrive on training disguised as play.

Dogs tend to tune out if they associate commands with an end of pleasure. That's why you should use a positive tone of voice and reward appropriate responses when you give commands such as "come" (when calling a dog away from other dogs) or "leave" (drop the item in your mouth) and even "no" (stop what you are doing and look at me because I can propose a more productive activity for you). And remember that, when issuing commands, consistency is critical, since it's the very best way of avoiding confusion in your dog. Once you've decided on a command for a certain behavior, it is generally best to stick with it. A combination of vocal and visual cues, such as verbal commands with simultaneous hand signals, can be confusing because it makes it hard to be consistent. Clearly, giving a command more than once in swift succession ("Sit! Sit! Sit!") is being inconsistent, since the dog might assume it must hear the command repeatedly before responding. Change a command only if the dog has been retrained to offer a different or better response. Giving a command and then not ensuring that it's followed is a particularly bad practice, because in effect, it trains the dog to ignore commands.

Think about how you use your dog's name before a command. Adding the name can ruin your attempts to be consistent and can confuse a dog that has just learned a new response to a particular command. The problem is that the name and the command together may sound very different from the command alone. We discuss the significance of names for dogs in chapter 7, in the section Do I have your attention? (page 119).

Ignoring commands

Confused dogs learn to avoid responding altogether. They learn to filter out the noise (including commands) coming from their humans because it is largely irrelevant to them; cats are especially good at this. Commands become irrelevant when they lose their association with predictable events. Moreover, when they lose their association with pleasant consequences or become associated with bad outcomes, why would they not be ignored? It is a rare dog that willingly subjects itself to unpleasant outcomes without at least trying to dodge them, even if only on an experimental basis. A dog's attempts to avoid unwanted events are sometimes labeled defiance, but they really represent a natural response to discomfort and are far from calculated.

Many owners set themselves up for failure when they disregard their dog's current motivation and issue a command that runs counter to what the dog is currently enjoying. The best example is calling your dog when it is clearly focused on running away to play with another dog. The called dog may experiment with ignoring the command and be rewarded by getting closer to its playmate. So, the outcome of ignoring the command is immediate pleasure. Effectively, the owner is training the dog to run away when it is called. In this context, there are important lessons to be learned from falcon trainers, who make it a rule never to *call in vain*. They religiously avoid calling a bird back to them when it is clearly focused on departure.

The key to success involves the use of exquisite timing. In the case of a recall command, the challenge is to time the command so that the animal is about to leave the object of its desire and thus to make it more likely that the instruction will be obeyed. If the dog has no intention of leaving the object, you're far better off approaching him and gently making a citizen's arrest that teaches him that crime (in this case, ignoring you) does not pay.

Ensure that your dog heeds commands by:

- using a single command for a single response
- issuing commands only once
- ensuring your dog responds to each command
- never calling when you are likely to be ignored.

Confusion and conflict

A dog that has competing needs or motivations can become conflicted. It's important to be aware of this because it's probably most distressing for the dog, and it can prompt him to try new responses to solve the conundrum. Clearly, if one of these new responses is rewarding, the behavior will be repeated in the future. Competing interests form the core of communication problems between humans and dogs. When an owner fails to understand what is motivating a dog to perform a certain behavior, it likely to be labeled as disobedience. The natural response is to repeat the command, usually with greater emphasis on a certain syllable.

Consider what happens when an owner leaves a young dog on a leash outside a shop. There are competing drives to follow his owner into the shop and to get some relief from the discomfort around his neck from the collar and leash. After a brief period of tugging against the leash, the dog establishes that he has a problem. He wants to get into the shop—but cannot; he wants to stop the neck discomfort—but cannot. It's been said that learning occurs when the only way to overcome a problem is to create a solution. Puzzles that haven't been encountered before require new solutions, especially if tried-and-tested methods don't work. This is where your dog begins to experiment. The dog sits—no reunion with owner. He lies down—no reunion with owner. He pulls in a new direction—no reunion with owner. He becomes distressed. He chews the leash, but it is made of chain and hurts his teeth—no reunion with owner. He barks and listens for any response from his pack. He barks and barks. It feels quite good. The owner comes out to tell him to be quiet. Bingo!— reunion with owner. Clearly, the next time the dog encounters restraint outside the shop, he will bark sooner (because of the previous reinforcement). The owner should ignore the racket, even though the dog will bark louder and longer (because of the *frustration effect* of extinction). Equally, he may generalize to make this response whenever he encounters behavioral conflict, when, caught between two conflicting motivations, he simply doesn't know what to do for the best. He has taken some formative steps to becoming a barker.

Conflict can also lead a dog to experiment with alternative responses: in other words, to come up with entirely new potential solutions. A dog that has been successful in testing new responses will certainly seem more creative. In addition, dogs are adept at differentiating between certain and uncertain situations, a skill that means they pay the costs of seeking extra information only when it is strictly necessary. This is when they are described as learning to learn, and this is when coaching becomes extremely rewarding because it allows you to draw out (educate) exactly what you want.

Inconsistency

I have been emphasizing in these chapters on learning and training that consistency is essential for rapid learning and effective training. So perhaps it's worth pausing to consider the many ways we are inconsistent. When dogs begin to ignore established commands, inconsistency is often the cause, and one of the chief sources of inconsistency is repetition. To your dog, a command repeated once or twice in swift succession is completely different from the original, single command. Furthermore, issuing a command repeatedly trains your dog to expect the command to be repeated. In a sense it means that each repeated command is nothing more than a warning that the real trigger will come along soon.

Hands-on, hands-off

After spending so long addressing the basics of dog training, I want to stress that the best trainers don't depend on physical restraint to get the best out of their dogs. Hands-off training begins by using your knowledge of the dog's body. For example, when the head goes up, the bum (being attached to a fairly rigid spine) dips down. So, raising one's hand as a visual cue for sit automatically sends the bum towards the ground. This is surely more elegant than reaching down with your hand and pushing the dog into a sitting position. As a hands-off trainer you have the luxury of being able to choose whether to use a leash; hands-on trainers have no choice.

Not having to touch your dog to get him to respond when close by means that he is much more likely to respond when at a distance. This doesn't mean that you should aim to get your dog off the leash as often as possible. Roads are always dangerous for dogs, but owners are always surprised when their dog is killed by a car. A well-trained dog walking calmly to heel is a joy to watch—he might not need to be on a leash, but if vehicles are within 500 yards, he is far safer.

Off-leash does not mean feral

The importance of aerobic exercise for a dog's health makes hands-off control compulsory for any city dog owner who wants to provide their dog with adequate exercise. The special merits of off-leash exercise have made people more aware of the importance of off-leash dog parks. We owe it to the next generation of dogs to manage these facilities thoughtfully. A well-designed dog park is at least an acre in area. Ideally, it offers a complex landscape with dips and knolls, trees and bushes, and has at least two entrances/exits, each with double gates for safety. Peer pressure within the park usually moderates the behavior of its human users, and this has a similar effect on their dogs. The Association of Pet Dog Trainers (APDT) has developed some thoughtful guidelines for the conduct of dog owners attending these parks.

USE OF LEASH-FREE PARKS

Do	Don't
Aim to use the park when it is reasonably quiet.	Don't enter if lots of dogs have gathered on the other side of the gateway. Very few dogs can cope with being greeted by a mob.
Leave dog toys at home to avoid them being guarded.	Don't allow your dog to guard you as a resource.

Do	Don't
Remove your dog if it bullies others.	Don't let dogs sort it out.
Remove your dog if it is being bullied by others.	Don't force frightened dogs to remain in the park.
Keep moving around the park so that your dog keeps an eye on you and does not develop transient territorial defense of a certain zone within the park.	Don't hang around in one spot, especially beside a gateway.

You shouldn't assume that your dog will enjoy being in a dog park, especially as he ages. And even if he does relish being there, never fall into the trap of going there every day, since this can cause him to become territorial. Go to other areas and keep your dog's sense of discovery keen.

As we've recognized, the best training opportunities come when all other dogs in the vicinity are under control. Too many owners switch off completely when their dogs are racing around the park. Some use the park as some sort of day-care facility and even leave and go shopping! It is up to responsible dog owners to spread the word about good dog-handling skills. Dog parks are an ideal forum for exchanging views on dog health, management and behavior. However, they are often truly brimming with self-appointed experts. Integrating APDT training clubs into dog parks is one of the best ways forward here, because it promotes a unified conduit for quality information and encourages basic dog training within the parks themselves.

CHOICE CUTS

- Dogs' value systems change with circumstances and also with time.

- Certain behaviors should be taught at certain times in a dog's life.

- Social learning cuts the high costs of trial-and-error learning.

- Competing interests form the core of communication problems between humans and dogs.

- Dogs thrive on training disguised as play.

- The best trainers never rely on physical restraint to get the best out of their dogs.

- Coaches who accentuate the positive end up with the most creative dogs.

In foxhounds, the use of a "hound-couple" brings about the transmission of trained responses from older hounds to younger ones by a process of negative reinforcement: Pressure on the collar disappears when the novice makes a correct response.

14
All Dogs Are Not Equal

Breed differences

In the process of domestication from wolf to dog there has been a reduction in relative brain size: Brain sizes in adult dogs are considerably smaller than in adult wolves of equivalent bodyweight. However, the differences between Uncle Wolf and Feral Cheryl and one domestic dog and another do not end there. Although all dogs share 99 percent of their genes in common, DNA studies of 414 dogs from 85 different breeds have been used to establish four basic breed categories:

- Ancient (including those breeds immediately derived from the Asian gray wolf, such as the Siberian husky, Alaskan malamute, shar pei, akita, Afghan hound and saluki)
- Hunting (including bloodhounds and golden retrievers)
- Herding (including collies and Belgian sheepdogs)
- Guard dogs (including mastiffs, bulldogs, boxers, Rottweilers and German shepherds)

It's thought that the three non-ancient groups emerged largely as a result of breeding programs during the past few hundred years.

Clearly breeds behave differently, especially in how they respond to potential prey. Terriers snap and shake; collies snap and chase; toy breeds yap, snap, stare, sulk and charm. In the grand, unwritten history of dog use through the ages, some hunting *Canidae* just showed tremendous patience and could trot and lope along after their quarry until it became tired. These reasonably crude generalists suited early human hunters, who favored similar techniques that revolved around persistence and an eventual coup de grâce.

It's easy to fall into the trap of imagining that all wolves hunt large prey only and that they therefore need to hunt in packs. On the contrary, many wolves forage on prey as meager as mice and are therefore independent of the pack in this respect. It's believed that individual foraging behavior of this type gets wolves through hard times without large prey. A dog's ability to forage and survive alone may be a significant legacy from Uncle Wolf.

Dogs in packs

The ability to gel as a group is the essential difference between so-called packing and non-packing breeds. Hounds bred to hunt in a group are obvious examples of breeds that work well as a pack, but bearded collies, old English sheepdogs and corgis also fall into that category. Packing breeds live together harmoniously, one member of a pack often taking a leadership and another often taking the role of caregiver. The distinction between packing and non-packing breeds is of interest to breeders more than behaviorists, who deal professionally with small groups of dogs or individuals but nonetheless tend to assume that all dogs want to behave as a pack because they have evolved that way. Are so-called packing breeds better at fitting into a large human family than, say, the mythical one-man-dog breeds (e.g., Chow Chow)? Breed differences in hierarchical organization and deference to status are certainly worth investigating, but, as we've seen, there is a danger of throwing the baby out with the bathwater if we assume that all breeds have an equal need for leaders or alphas or, for that matter, life-coaches, caregivers and companions.

Sight-hounds and scent-hounds

One of the most important delineations among hunting breeds is between the sight-hounds (examples of which appear in the *Ancient* category above) and the scent-hounds (examples of which appear in the *Hunting* category).

SIGHT-HOUNDS

The salukis, greyhounds, whippets and Afghans are all sight-hounds and they all have long noses. This certainly isn't to help them smell their prey. They may be pointy-headed, but they are also generally long-limbed, and perhaps this means that they are generally long-boned. It's not completely clear whether the long bones needed for catching prey are likely to mean that the skull is automatically going to be longer than that of a non-chasing breed. The sight-hounds also tend to be sensitive rather than clumsy. Once they've matured, all members of the sight-hound breeds are aloof rather than boisterous and ebullient. Rather than looking at you, they have a reputation for looking through you, or at least beyond you. They scan their visual field, looking for relevant stimuli, and that, for them, means anything chase-worthy, anything prey-sized and moving.

We've already seen how long noses are associated with good peripheral vision, whereas dogs with short noses are able to see tremendous detail in the center of their visual fields. Dogs with long skulls also tend to have almond-shaped, rather than round, eyes, but it's not clear how this may affect eyesight.

SCENT-HOUNDS

Scent-hounds typically have their heads to the ground as they track odor gradients and sample fresh incoming scents. Their chief skill is to distinguish among significant targets and background odors. By seeking out the important odor gradients, they can find the source of the scent. Their olfactory apparatus within the skull is well developed, but scent-hounds don't necessarily have big noses. The membranes used to pick up odor molecules are folded over on themselves in moist, intricate layers that, if ironed out, would cover a soccer field.

It is said that bassets and bloodhounds have long ears to disturb ground-based odors and thus make them easier to detect. I am skeptical about this, since the ears hit any scent after the nose has passed them, so their effect on the movement of odor molecules would seem to be irrelevant. There is virtually no need for a scent-hound to worry about what is going on behind it. In contrast, the function of the ear-flaps as a means of blocking out redundant noise, such as that made by the dog itself, may be significant. The dog that crashes around in the undergrowth may be searching for injured prey or simply flushing out animals attempting to hide. So its attention is generally on objects (obstacles to be traversed or investigated in case they are hiding game and, of course, the game itself).

Individual differences

Recent studies have focused on the way dogs adapt to social change, suggesting that during the process of domestication, dogs display a set of socio-cognitive abilities that let them communicate with humans in unique ways. Additionally, *Canis familiaris* shows unique morphological diversity among breeds that make it a particularly interesting research model when investigating both canine and human genomes. These exciting avenues of research are showing how human and canine research can assist one another. Obsessive compulsive disorder is a terrific example, with the suggestion that tail-chasing dogs are ideal models for research that may help humans afflicted with repetitive tendencies, such as compulsive hand-washing.

In many ways, veterinary behavioral medicine has developed along similar lines to the human psychiatric field, but, as a relative latecomer, it can capitalize on recent advances in psychobiology and learn important lessons from human psychiatry. Whereas information from dog studies has traditionally informed human psychology, there is now a chance for dogs to benefit from human studies. One area in which this may be particularly relevant involves the study of individual differences in positive motivation and personality. Emotionality, fearfulness, boldness, impulsivity have all been used to describe the dog's approach to life. They may reflect qualities that owners recognize as loyalty, bravery, foolhardiness, willingness-to-please and even a sense of humor.

The features of behavior that make one dog different from the next are essentially what's known as personality in humans. Research in this domain is tremendously important because it's been suggested that dogs at either end of the spectrum of personality types may be prone to inappropriate and problem behavior and thus be regarded as having "clinical mental illness." This is the focus of a recent study by Jacqui Ley, an Australian veterinarian. She asked more than 1,000 owners to rate their companion dogs on more than 60 personality adjectives. Statistical analysis identified five underlying characteristics that could be used to cluster the owners' responses into the following categories: extraversion, neuroticism (or cautiousness), self-assuredness/motivation, training focus and amicability. These are remarkably similar to the so-called Big Five personality types described in humans (extraversion, neuroticism, openness, conscientiousness, and agreeableness). It will be interesting to see how successful these five labels are in describing the traits of dogs as scored by behavioral tests and how stable this system of labeling is over the life of a dog. Taken together these steps will facilitate the development of tests that predict the suitability of puppies for certain tasks and companion homes. This sort of inquiry offers considerable hope for improved matching of dogs with owners and will allow owners to be more definitive when describing the dog traits that they truly value. Breed differences in the distribution of these five traits will also be worth investigating, since potential puppy purchasers must currently rely on breeders' descriptions of breed tendencies—largely drawn directly from written breed standards—even though temperament is scarcely scored at all in the show ring, where most breeding lines are selected.

The importance of temperament

Even though the selective pressures humans have applied to dogs over the centuries have changed, we've been steadily filtering out genes that weren't useful to us. So what became of all those dogs that didn't meet the mark? Traditionally, they would have been discarded, abandoned, eaten or simply culled. It is interesting to think of the road to domestication as being peppered with checkpoints.

The earliest proto-dogs were those of Uncle Wolf's buddies that could tolerate being close to humans to forage in their waste. Of course, any of these ancestors unable to adapt to the human household have been removed from the gene pool, especially if they were inappropriately aggressive. They have been stopped by the checkpoints. In contrast, pups from dams and sires that were well liked in a community were desirable and therefore in demand. They received gold passes to cross the checkpoints.

These days, responsible pet owners are encouraged to neuter even the most well-adapted companion dogs. While this has brought about a welcome reduction in unwanted canine pregnancies, it has also meant that a greater proportion of dogs in our community are the product of specialized dog breeders who don't prioritize the selection of breeding stock on the basis of their adaptability and social skills. Instead, they select for characteristics of appearance that succeed in the show ring.

So, to generalize, we now have something of a mismatch: Humans who are less experienced at interacting with, training and managing dogs than their forebears and from whom society expects more, are being matched with dogs bred primarily for the show ring. The clash of these cultures undoubtedly makes it all too easy for the media to vilify dogs. This, in turn, makes fearful people phobic. And, to generalize further, when the media wants to vilify dogs, it targets the so-called fighting breeds because they are a reasonably soft target, being so often owned by the least responsible owners. In essence, the public's general inability to avoid, manage and respond to dog aggression has caused several countries to resort to breed-specific legislation intended to control ownership of some dog breeds.

Temperament differences reported by caregivers

Psychologists assess a person's responsiveness to rewarding and negative experiences (their positive and negative activation) with a mind-measuring (psychometric) interview. Although we can't interview nonhuman animals, tools have recently been developed to clinically evaluate patients in veterinary behavioral medicine and to create psychometric profiling using their owners, caregivers and handlers. Studying a dog's responsiveness in this way has two benefits. First, these psychometric tools may help to identify dogs that are at high risk of developing fears, phobias and anxiety problems. Second, they may detect those that are especially sensitive. This is critical because, despite what trainers may claim, even if they don't use physical punishment, most use a combination of reward and punishment. (Remember that, by definition, punishment is anything that reduces the frequency or probability of a response in the future—so any deterrent, even a verbal reprimand, can act as a punisher.)

Given that all dogs normally encounter some form of training in their life, this sort of profiling will allow a training style to be customized for each dog. For example, a dog that is particularly sensitive to both positive and negative outcomes could be trained with an absolute minimum of negative outcomes (which, in this context, could include something as apparently innocuous as disappointment).

Assessing the behavior of dogs along the continuum from cautiousness (also labeled shyness) to boldness may help predict levels of performance in working dog trials. Rather than being relevant only to working dog enthusiasts, this research can help to characterize companion animals, especially those that are patients of veterinary behavior therapists. As any therapist will acknowledge, even when owners comply with the prescribed steps in a behavior-modification program, some patients respond better than others. At least some of this variation may be explained by differences in the emotional sensitivity of the dogs in question.

Temperament testing

Despite greater controls on dog ownership in some countries, many behavioral problems persist, including aggression. Meanwhile, the dogs often pay a high price. Many thousands of adult dogs are relinquished to animal shelters or abandoned in the community each year. Before being rescued, these dogs may have endured starvation, injury, abuse, illness, isolation stress and disruption of attachment relationships. Approximately 30 percent of the dogs unfortunate enough to enter a welfare shelter are euthanized. This not only represents a potential welfare issue for the dogs concerned but can have psychological ramifications for the people whose job it is to temperament test and kill large numbers of dogs every day.

Numerous educational strategies have been developed to improve the way in which humans, especially children, care for, train or interact with dogs. Meanwhile, to prepare dogs for the demands of modern urban existence, dog socialization classes (for both puppies and adults) have become common. Unfortunately, these strategies don't guarantee that the dogs made available to prospective owners by breeders and rescue shelters are going to make happy, healthy companions.

Breed-specific legislation

In an attempt to outlaw powerful dogs originally bred for fighting, breed-specific legislation has been introduced in many countries, but it is very controversial, not least because so many of the dogs targeted are really types (dogs of certain shape) rather than (formally registered) breeds. Opponents argue that the subject of legislative control should be "deeds not breeds." But this implies that we have to wait for dogs to bite people before we can even identify them as a risk and begin to manage them. Biting aside, any focus on deeds demands that we first agree on what dog behaviors are acceptable or not to the public and, second, agree on how to assess them.

OWNER EXPECTATIONS

There hasn't been much research on owner expectations of their companion dogs. Those who surrender or abandon their dog don't tolerate behaviors that many dog owners accept. So, in this sense, the protocols at the domestication checkpoints are inconsistent. The most commonly cited reasons for relinquishing adult dogs to shelters are incompatibility with existing pets, aggression, separation-related distress, hyperactivity, boisterousness and a tendency to escape. I would argue that three of these problems (the last three in my list) can, at least in part, be addressed by regular exercise that tires the dog. Perhaps we should be doing more to ensure that dogs can be exercised and that, before acquiring dogs, owners understand what adequate exercise involves. It's possible also that owners are understating the significance of the problems for fear that if they tell the truth, the dog might automatically be destroyed on safety grounds.

It should be a priority to better match dogs with owners and their expectations. This relies on some behavioral assessment of individual dogs, a challenge that is currently without a solution. Attempts have included the development of puppy temperament tests to reliably predict adult behavior. Unfortunately, these have had only limited success, mainly because the range of behaviors expressed by puppies before being placed in new homes is so limited.

Reputable dog-sellers usually include some behavioral characteristics when describing available dogs, but such recommendations to prospective owners are typically based on breed standards, so they tend to describe characteristics of the breed rather than those of individual dogs. Fundamentally, this sort of advice on dog care is often flawed because it is largely informed by the original functions of breeds—for example, hunting, working, livestock guarding or herding—functions that are no longer relevant to most dog owners. Equally, as we've noted, existing breed standards aren't very useful in describing modern dogs. According to a recent Swedish study, rather than reflecting traditional breed functions, many popular breeds are characterized by their effectiveness in the show ring. The Australian Small Animal Veterinary Association (ASAVA) recently recommended introducing temperament testing in dog shows when it announced the ASAVA Temperament Prizes to be awarded at all the major dog shows in Australia. At the time of writing, the temperament test most likely to be applied will be the Delta Society's Canine Good Citizen Test. This involves testing a dog's reaction to a selection of challenges drawn from a pool of seven types of distraction: a person on crutches, in a wheelchair, or using a walker; a sudden closing or opening of a door; dropping of a large book; a jogger; good-natured pushing and shoving or animated excited talk and back-slapping by persons; a person with a shopping cart; or a person on a bicycle.

Rescue shelters

Today's most rigorous domestication checkpoints are the rescue shelters, where dog behavior is assessed. Nearly all shelters conduct some form of health and behavioral assessment before making dogs available for adoption. Dogs that fail these tests are often euthanized, an appropriate outcome because routine re-homing of animals with significant health or behavioral issues wouldn't be acceptable to those of us who adopt dogs from shelters. That said, it is also unacceptable to kill dogs that have fallen victim to an arbitrary test or a lackadaisical tester.

The chief problem with current rescue shelter practice is that their behavioral testing isn't consistent. Many shelters currently run tests developed in-house that mainly measure how well a dog reacts to veterinary handling. I like this test because it helps to reveal whether a dog is likely to bite a veterinarian, but then again, I have a vested interest. In fairness, these tests usually overlook factors that may predict whether or not a dog could be successfully integrated into a new home. This means that shelter dogs failing a behavioral test may be euthanized unnecessarily, while dogs from another shelter may be re-homed after passing a test that fails to detect whether they are genuinely suitable for re-homing. Such inappropriate tests probably explain why from 5 to 20 percent of dogs adopted from shelters are returned to the shelter shortly afterwards. Aside from the legal implications of selling risky dogs, this is a problem because it compromises the faith potential adopters have in the shelters and is probably distressing for dogs that are serially adopted. It also wastes shelter resources.

There is an overwhelming need for a validated test suitable for use by rescue, breeding, training and selection organizations in assessing adult dogs. That said, such a test needs to be calibrated against the general population. This is possibly the single most-important step in any behavior assessment exercise, and so far it has been overlooked. While it's all very well to assess the population of dogs in rescue shelters, the results are essentially meaningless until we relate the results from such an exercise to the results of the same assessment protocol being applied to the general, owned-and-loved dog population. Only then can we see what the tests mean, and we may be in for a bit of a shock. We may discover that the general public is tolerant of the same suite of unwelcome traits and a similar range of extremes in temperament. This will add great weight to the need for profiling potential owners, since it would clearly suggest that surrendered dogs are not necessarily dripping with issues and that it is owners' expectations that are often at odds with reality.

Given that inherited factors underpin some key individual differences in temperament, such as anxiety, the right assessment tool for profiling dogs should make dogs safer and easier to live with. This will mean that dogs selected for work, companionship and breeding will be suited for the demands currently placed on them.

Laterality (left- and right-handedness)

Recent studies have exposed intriguing links between laterality and a dog's temperament. The influence of brain lateralization on other characteristics is equally fascinating. Professor Lesley Rogers at the University of New England, NSW, has reported evidence for behavioral differences in common marmosets that were linked with hand preference and brain-hemisphere lateralization. Right-handed (left-hemisphere-dominant) marmosets entered a novel room faster and touched more of the novel objects than did their left-handed (right-hemisphere-dominant) counterparts. This led to the conclusion that left-handedness was associated with fearfulness while right-handedness was associated with exploration and approach.

CHEW ON THIS

It was originally thought that lateralization was a quality that appeared only in humans, but this has already been convincingly discounted. Indeed, it seems that there is a basic pattern of lateralization common to all vertebrates, including humans, nonhuman primates, dogs, amphibians, cats, whales and rodents, as well as birds, fish and reptiles. The wide range of lateralized animals identified today implies that modern species have inherited brain lateralization from a common ancestor.

A recent Australian study by Nick Branson, studying at the University of New England, reported evidence of lateralization in dogs by showing paw preference when holding a Kong (a rubber dog toy) containing food. This study showed the first links between canine anxiety (in the form of thunderstorm phobia) and lateralization. This may prove to be of critical importance. Intriguingly, the dogs that showed no preference for either the left or right paw were more likely to be fearful of loud noises. If such dogs are consistently more fearful and suspicious, then guide dog associations and indeed many other service dog users, such as the police, quarantine services and army, may choose not to select such animals as part of their recruitment process to avoid wasting training resources on them.

Sex differences in laterality have been reported in many species, with males more likely to be left-handed than females (for example, in lemurs, chimpanzees and horses). It's been suggested that testosterone produced by male embryos during pregnancy (and potentially affecting female littermates) could account for such sex-related differences. Assessment of lateralization in 53 dogs of various breeds through paw preference tests indicated that females tend to prefer their right forepaw, while males favor their left forepaw in the same tasks. A subsequent study showed the same motor bias in males versus females. This is significant, since dogs of both sexes are routinely recruited for guide dog work.

LATERALITY IN GUIDE DOGS

This is an especially important attribute since these dogs are required to walk exclusively on the left-hand side of their handler. It may be that this convention favors dogs with sensory laterality dominated by the right hemisphere, for example, because this facilitates observation of the visual environment by the left eye. This global dog-handling convention is also rigorously adhered to in police dog training. Once working dog trainers know more about the strengths and limitations of left- and right-preferring dogs, they will need to train fewer dogs because they can make more-informed selections at the outset of training. At the same time, having a significant motor preference may make a dog generally less able to work on a given side of the handler's body, so the study may challenge the world-wide left-of-handler convention and allow dogs with whatever bias to be used more effectively. Of course, we also need to estimate the effect of bias coming from strongly lateralized handlers.

Aggression toward other dogs is why many trainee guide dogs and, to a lesser extent, trainee police dogs fail their eligibility tests. Significantly, a recent study showed a relationship between laterality and aggression, with right-pawed animals being more aggressive. In contrast to dog–dog aggression, aggression to humans as manifested by a so-called "prey drive" is highly desirable in general-purpose police dogs. The necessary research may allow this form of aggression to be measured and even predicted.

HOW TO ASSESS WHETHER YOUR DOG IS LEFT- OR RIGHT-PAWED

What you need:

- Some soft dog food—your dog's regular brand is less likely to cause a stomach upset than anything else
- A Kong toy
- A sheet of paper numbered from 1 to 100
- A pencil
- Plenty of time. This test can take up to four hours.

Method:

1. Pack the Kong with food and place it in front and to the center of your dog's paws.

2. Record the paw used to first touch the Kong at number 1 on the recording sheet as left paw (L) or right paw (R). Definitions for the way in which interactions are classified as L or R appear below (interactions with both paws placed separately on the Kong at the same time are not counted).

Left paw (L)	Right paw (R)
- Left paw on Kong, right paw not.	- Right paw on Kong, left paw not.
- Left paw over right paw on Kong.	- Right paw over left paw on Kong.
- Left paw on top of Kong, right paw underneath.	- Right paw on top of Kong, left paw underneath.

3. Continue to record the paw used to touch the Kong until the dog has made 100 paw interactions classified as left paw (L) or right paw (R).

4. If the dog repositions its paw or paws on the Kong, without the paw or paws completely leaving the Kong, that interaction with the apparatus is not counted or recorded.

5. If the dog holds the Kong down with a certain paw or with both paws for longer than 10 seconds, gently remove the Kong from under the paw or paws and place it in front and to the center of your dog's paws again.

6. Some dogs retrieve only the top layer of food from the Kong. For these dogs, the Kong must be topped up with food to achieve the 100 paw interactions. Large hungry dogs empty the Kong quickly so they will also need to have it topped up.

INTERPRETING THE RESULTS

Once you have recorded 100 interactions of the left or right paw with the Kong you have sufficient data to determine whether your dog is left-pawed or right-pawed. Dogs that use their left paw 64 times or more are left-pawed. Those that use their right paw 64 times or more are right-pawed. Dogs with less than 64 uses of either paw are considered ambidextrous.

Breaking news

Through studying a dog's laterality, we hope to be able to confirm that certain individual dogs can be detected as having a significant risk of noise phobia and separation-related distress. Noise-phobic dogs make poor candidates for general-purpose police work, since the problem may cause distraction and compromise performance in dogs working under noisy conditions, such as sieges and riots. Meanwhile, separation-related distress is very common in the current working guide dog population, a phenomenon that is a worry to guide dog associations worldwide, since it contributes to concerns about the welfare of these dogs. It is critical that this research is completed before more inappropriate dogs are selected for guide and police work. When applied to contexts other than that of working dogs, screening for individual differences, including emotionality and laterality, could help by identifying dogs vulnerable to fear triggers. These could then be habituated more carefully before being made available as companion animals.

CHOICE CUTS

- All dogs share 99 percent of their genes in common, but their morphological diversity is phenomenal.

- There are four breed categories: ancient, hunting, herding and guard dogs.

- Before the emergence of dog shows, breeding focused on behavior rather than appearance.

- As a result of domestication, dogs show a set of socio-cognitive abilities that let them communicate with humans in unique ways.

- Extraversion, neuroticism (or cautiousness), self-assuredness, training focus and amicability are all used to describe dog "personalities."

- Inherited factors explain some key temperamental differences among dogs. Profiling dogs will help us select those that are more likely to fulfill our expectations.

- There are fascinating links between laterality (left- and right-handedness) and a dog's temperament.

Left and right paw preferences offer fascinating insights into the dominance of right and left brain hemispheres respectively. In other species, brain hemisphere dominance has been used to explain profound differences in personality.

15
Working Alliances

At the shepherd's bidding, his sheepdog tears off in search of the flock that needs sorting out. The police dog is unflinching in his search for that lost child, and the livestock-guarding dog remains devotedly with his charges. Many of the dogs in these examples receive no food rewards to train them or keep them working. Some trainers of working dogs even espouse the "Treat 'em, mean; keep 'em keen" philosophy. Pet owners would do well to heed the lessons learned from the hours, days and years humans have spent working with dogs. Of all the species humans have recruited to help them at work, dogs are the most ubiquitous. They are sociable problem-solvers with a quality that trainers and owners often refer to as "a will to please."

We might be tempted to interpret dog work as willingness to meet our needs, but let's consider this in the light of what we know from the legacy of Uncle Wolf by looking at the questions: How can a dog be sure when other members of its pack are feeling pleased or otherwise? Why would willingness to please others be advanced by natural selection, where the emphasis is on success and competitors of the same species are close by?

Canine workaholics

Good breeding for behavioral characteristics and strategic channeling of their motivation explains why dogs can be such willing
workers. As social animals, dogs work well as members of a team,
not least because they learn from those around them. Furthermore,
they are sensitive to social reinforcement, such as attention, and to
unpleasant experiences, such as being ignored or isolated. In other
words, rather than wanting to please their trainers, dogs may simply
enjoy their work as well as the company and the verbal praise of their
human coaches. For some breeds, in the right context, a reward can
be being allowed to carry on working, while a punishment might be
being stopped from working.

The dog's many impressive and exploitable qualities include an
outstanding sense of smell. So we find working dogs that use their
noses to detect odors from explosives, truffles, termites and even the
scent of broken grass for competitive tracking trials. Also, their charac-
teristically high motivation to play can be redirected towards humans—
either dead or alive—for police and search and rescue (SAR) work.

Training guide dogs involves rewards for not doing certain things
as well as for doing others. For example, the dog must be trained to
cross a road only when the way is clear, regardless of any instructions
from the owner. If a friendly-looking puppy on the other side of the
road makes enticingly playful overtures in the direction of the guide
dog, the innate motivation to play must be over-ridden by the motivation to behave according to prior training.

Working dog welfare

We fondly imagine that working dogs have a different work ethic than
pets and that they somehow have a warm glow of satisfaction at the end
of each working day. Even the hardest-nosed manager realizes that the
health and welfare of working dogs can enhance the length of their

working lives, but it's worth also looking at some of the drawbacks of a working dog's life. For each working dog situation (see the table below), training, feeding, housing, and exercise can compromise dog welfare.

Jobs for dogs

Type	Purpose
Farm dogs	Sheep herding Cattle herding Livestock protection
Hunting dogs	Retrieving Pig hunting Pack pursuit/tracking Underground work
Government dogs	
Police	General purpose (tracking/apprehending/crowd control) Firearm and explosives detection Drug detection
Military working dogs	Army Air force
Customs (detection of illegal goods such as drugs & weapons)	Multipurpose response Passive response Active response
Quarantine (detection of exotic pests & diseases)	Active response Passive response
Correctional Services	General purpose Drug detection
Service dogs	
Assistance	For physically impaired
Guide/seeing eye	For visually impaired
Hearing	For hearing impaired
Pets as therapy	Visiting or permanent placement
Search and rescue	Volunteer dogs/handlers able to assist Police/Emergency services
Racing dogs	Greyhound Sled

Private industry		
Security/Guard	Working mainly with handler	
	Working mainly without handler	
Detector dogs	Explosives detection	
	Drug detection	
	Truffle detection	
	Seizure alert	
	Termite detection	

Let's look at the training methods used for each type of work. There is a difference between training a dog to express an innate tendency and training a dog to offer an entirely new response. Although these days, professional dog trainers are sharing a lot of training tips with pet dog trainers, tradition is holding some of them back big time. The persistence of choke chains in guide dog training is a good example. And a recent case in the United Kingdom of a police dog handler being charged with cruelty for hauling his dog in with a choke chain is another.

Welfare can be compromised by harsh, coercive and punishment-based training techniques that may involve the use of choke chains and, not uncommonly, electric shock collars. In the light of modern learning theory, many time-honored methods of dog training seem crude and inhumane. Take, for example, some traditional approaches to training for "man-work," an element of police dog training (in which the public also sometimes engages for competitive working dog trials). In order for the handler to demonstrate that the dog is taking hold of its human target with conviction, traditional training involves habituating the dog to being bludgeoned about the head by the "victim." Indeed, until fairly recently these beatings were meted out not just during training but as part of the competitive trials themselves! To win, a dog would have to withstand repeated blows to the head. There are even reports that the more bloodletting the bludgeoned dog showed, the more impressed were the judges.

Of course, training dogs for work doesn't rely solely on punishers. Some of the most elegant uses of primary reinforcers occur while training sniffer dogs. Yet some may even criticize the use of rewards when, as part of the training program, food is restricted to ensure adequate appetite.

Dogs used for detecting explosives are a good example of animals that have been trained to link work directly with food. The element of classical conditioning in this training program is that the dog associates the explosive with food. This evokes salivation, which, being visible, can be rewarded. This physiological response is replaced by a behavioral one when the dogs learn to sit (a passive response; as listed in the table on page 256) to receive part of their daily food ration if they pick up the correct scents. Clearly, dogs working in the vicinity of possible explosive materials should never be trained to dig at the target of their interest. However, drug and food detection dogs in Customs policing contexts are often trained to make this emphatic response (active; as listed in the table above). Ultimately, these dogs don't get fed unless they have detected the tell-tale odor and responded appropriately. For this reason, handlers have to run the dogs through their paces even when they are off-duty on weekends and feed them their daily ration in portions as they respond correctly to odors the handlers have planted.

Motivators and reinforcers

In their early training, competitive tracker dogs usually learn that tracing the smell of broken vegetation is associated with primary reinforcers, such as food. Often the same olfactory cues are then linked to toys that have been hidden. So, using the innate skills that helped Uncle Wolf track and harry prey, these dogs eventually locate conditioned stimuli that provide them with fun, not food. Similarly, when foxhounds work as a pack, they do so not for food reinforcers but because they enjoy being part of a team and chasing prey.

When we give dogs jobs that recall predatory behavior, it seems to be the excitement of hunting that keeps them keen. Dogs looking for toys that have been associated with certain scents or visual stimuli seem to relish locating and securing these items because they represent their prey. Some working dog trainers prefer to use toys as a reinforcer because they reckon that feeding routines may interfere with work. This seems especially valid for police attack dogs, because such dogs can be called to arms at any time of day. A food-driven dog would be of little use right after it had eaten its evening meal.

There are considerable differences in the stimuli used in various types of training. Search and rescue dogs are trained to track airborne scents (rather than ground tracking), which allow them to detect scents up to a little more than a half mile away, wind permitting. By contrast, police siege dogs respond to visual stimuli. For them, the sight of a human running away holds the promise of rare fun indeed, because they have been rewarded for playing with the jute arm sleeve "toy" worn by all fleeing "stooge criminals."

Although guide dogs are trained in traditional obedience exercises that generally have the humans calling the shots, this relationship has to change when the dog guides the human. When wearing the harness, the dog must have the confidence to know when to ignore commands from the human that are unhelpful. A similar sense of independence is also necessary for livestock herding, especially in competitive trials, since dogs first need to respond predictably, rapidly and consistently to approximately 60 audible commands while retaining some autonomy. One of the main reasons for this autonomy relates to the use of verbal and whistle commands that necessarily create a delay, since the dog doesn't hear the sound until it has traveled from the handler's lips to the dog's ear. During this delay the dog may have to act independently.

Feeding

Most working dogs receive adequate food, but they may not get to chew often enough. Furthermore, some sniffer dogs that have been trained to expect portions of their daily ration only once they've made a correct response, for example, sitting in the presence of a target odor, are kept hungry so that they are driven to work. Hunger is certainly unpleasant, but the debate about whether this technique represents a welfare issue is generally fueled by those who assume that lean dogs are unhealthy. Most veterinarians agree that being obese compromises health and welfare more than being marginally underweight. Indeed, recent studies have shown that feeding dogs 25 percent less than manufacturer's recommendations can significantly increase their lifespan. (This, in turn, suggests that manufacturers should urgently review their portion recommendations.)

Also, even though dogs thrive on routine, trainers sometimes deliberately avoid regular feeding times to prevent dogs from becoming too set in a routine and therefore being disinclined to work at certain potentially critical times of day or night.

Housing

Unless working dogs are seen as part of the PR wing of a corporation as, for instance, guide dogs are, the public is rarely invited to inspect the kenneling facilities for working dogs. So, while standards of hygiene are generally high (since cutting corners in this domain can compromise productivity), access to toys and playmates is often limited. Indeed, some managers of working dog facilities worry that enriching their environment potentially interferes with the dogs' work output. This philosophy goes against the latest evidence for laboratory rodents, which shows that enriched environments enhance learning and trainability.

Exercise

Dog work usually requires and implies locomotion. So most working dogs get more exercise than most pet dogs. However, the exercise some of them get at work can be quite regimented and formal. Working dogs thrive on routine, not least because territorial surveillance is part of a normal day for any dog, so trips with the handler can resemble this activity. The only snag is that, unlike the tour Feral Cheryl takes daily through her domain, working dogs do not get to revisit areas. The constant challenge of visiting new environments may prompt excessive marking and possibly diarrhea, most notably in entire male dogs (and police dog squads include many of these).

Clearly, play is very important to adult dogs, a feature that distinguishes them from adult wolves. It's interesting to speculate whether dogs that receive plenty of formal exercise actually miss frivolous play. Fortunately, as we have seen, many dogs are trained to work for toys (e.g., some sniffer dogs). Guide dogs, on the other hand, are sometimes perceived as having little fun while on the job. They regularly display an inhibited demeanor that may represent the legacy of the aversive techniques designed to reduce their response to distractions: other dogs (potential playmates); moving objects (balls to be retrieved); and food (dropped pies to be consumed).

Finally, exercise for some working dogs can be erratic and rather seasonal. For example, sheepdogs are often kenneled for extended periods (months, in some cases) when the sheep on their farm don't need to be moved. This brings tremendous concern for their welfare. Often chained, these dogs live in close proximity to large volumes of their own waste and the attendant flies these attract in summer.

Lessons from working dogs

We all marvel at the abilities and the keenness of working dogs, so let's explore their training and management and see what lessons we can learn and apply to companion dogs. For those of us not training dogs in working situations, it's easy to assume that the regimented environment of a working dog kennel is a prerequisite for getting the same exacting standards of performance from our dogs. It is true that kenneled dogs know the difference between being inside and outside a kennel and that this makes them especially sharp, keen and responsive when they are the focus of the trainer's attention. But this should not mean that unkenneled dogs are less trainable. Nor does it mean that some of the best features of working dog training can't be integrated into our own nonprofessional training efforts.

When the dog is with us but not undergoing formal training, it can still receive rewards for improved behavior. I call this *training by stealth*. Essentially, it means that the dog is constantly on the lookout for opportunities to score rewards unless it has been actively told to relax. This method owes a lot to what I have seen of working dogs in action.

The use of cues that tell dogs when they are not required to be completely attentive is critical here. A cue that tells the dog to stop concentrating is one of the most useful devices in the toolkit of top trainers. It means that words that sound like commands can be ignored and that rewards are not going to be missed. Essentially, it is an off-switch for the dog's attention and can be likened to the "At ease!" command comfortingly bawled by sergeant majors on parade grounds. Its strength lies in its ability to retain the dog's attention for when it is really needed, whether during training, work or competition. "All gone!" is a good example: It tells the dog that there are no more rewards currently available and that it can therefore stop offering the behavior for which it was last rewarded.

Dogs are learning all the time, but they don't have a sense of formal training or even competition. Such concepts are irrelevant to dogs, since they have evolved as opportunists and therefore need to be able to exploit all situations equally and to the utmost. Having said that, they do seem to be very good at detecting when handlers or even an audience are especially focused on them—as during training sessions, competitions and displays. There are certainly dogs that rise to the occasion with excellent, focused performances, while others seem to play to the gallery, milking their audience for reactions to almost comedic deviations from the trained script.

Given that dogs are such good learners, whether we mean to or not, we are constantly training them. Owners regularly tell me that they have not trained any unwelcome behaviors, but the truth is they have unwittingly trained almost all of them. In that sense you get the dog you asked for, even though you didn't ask specifically. Just as learning opportunities can creep up on us, they can also creep up on dogs, and these often provide the most effective lessons.

Rewards can come from nowhere but should preferably be accompanied by a click to confirm that they were linked to the preceding behavior rather than being completely providential. Breaking down the barriers between formal and informal training makes the dog less context-specific and ultimately more reliable under more circumstances.

CHOICE CUTS

- Pet owners would do well to learn from those who spend hours, days and years working with dogs.

- As social animals, dogs work well as members of a team, not least because they learn from each other.

- For some breeds, a reward can take the form of being allowed to carry on working, while a punishment might be being stopped from working.

- Optimal health and welfare of working dogs lengthens their working lives.

- Enriched environments enhance learning and trainability.

- Some working dog trainers favor toys as rewards for their dogs.

- You get the dog you asked for, even though you didn't ask specifically.

In many cases, working dogs that are reared, socialized and exercised as a group show how optimal training can help get the best out of life in the human domain. (Photo courtesy of Norm Keast.)

16
The Next Generation

Most of the dog breeders I know are well intentioned, passionate about improving the quality and fame of their own bloodlines, and clearly committed to promoting their breed and to the advancement of dogs in general. They live and breathe dogs. They spend days traveling to attend shows and, to get the results they want, they have to groom and exercise their dogs much more than most owners. They study breed characteristics, explore the merits of test matings and grapple with the complexities of veterinary genetics. And it is true they rarely make much money from breeding *per se*. The time, effort and expense of producing a high-quality litter are somewhere between considerable and tremendous. But there are often important differences between what a breeder and what I, as a veterinarian, might regard as a high-quality litter.

Greedy backyard breeders

It pays to ask why pedigree dogs cost as much as they do. And are they worth the price? Pedigree pups are expensive, but the better breeders can justify the price per pup in terms of the costs of feeding and housing the parents; stud fees; the entry fees, transport and accommodation costs of showing; the expense of importing new breeding stock; freezing semen; and so on. The only way to make a fast buck out of dogs is to cash in on someone else's efforts and to breed for quantity rather than quality. So-called backyard breeders are those keen to produce numbers of pups of a given breed, regardless of the parents' merits. They are often also the unspeakable characters who breed fast, never missing a season, never giving a bitch a break (let alone exercise), a bath or a bone to chew. They breed anything that vaguely looks like a representative of the breed without regard for the health of the pups and the welfare of the bitch.

One might suppose that, as with other purchases, a newly acquired puppy can be returned to the vendor if a serious problem emerges. But parting with a questionable dog is not as easy as parting with a broken washing machine. Dogs rapidly assume their place as members of the household or family, especially if they are allowed indoors. Pups and their new owners generally form a superglue bond with lightning speed. Indeed, this is why so few breeders ever have to take back the defective pups they should never have sold.

Despite the best efforts of professional dog breeders and dog-breeding organizations, there's very little pressure to do the right thing in dog breeding. Even if quantity is not the main focus of dog-breeding efforts, quality may suffer. Quite simply, backyard breeders are interested solely in cash, regardless of individual dog welfare. They are not interested in schemes to eradicate inherited disorders and so may provide a pool of unwelcome genes. Their pups may be more likely to succumb to inherited disorders than those of informed and committed breed enthusiasts who put money and effort into detecting and eradicating inherited problems.

Even without any financial incentives, another form of indiscriminate breeding is when regular pet owners breed from their favorite pet without really considering whether they are passing on unwelcome traits. Although numerically less of a problem than the products of unplanned matings and the wanderings of un-neutered dogs, such sentimental breeding can create pups that are difficult to re-home and so contribute to the problem of unwanted pets. In a sense, this is an argument for compulsory neutering of dogs in pounds; they have failed the basic test of compatibility—remaining in one home for life. Admittedly this test is a radical one. It throws dogs out of the breeding pool without establishing why they were abandoned or surrendered. Given that I was once asked, as a veterinarian, to euthanize a dog because it was the wrong color for the owner's new carpet (I didn't, of course), I am entirely unconvinced that all dogs are given the benefit of the doubt before being surrendered.

How much is that doggy in the window?

Sadly, far too many puppy purchasers don't care where their pups come from. This is perhaps because puppies are so universally appealing (we all know how well they sell toilet paper). This means that the high standards of care displayed by breeders are rarely appreciated, much less rewarded. The average potential puppy purchaser becomes weak at the knees when visiting breeding kennels and just wants to see puppies, puppies and more puppies. Effectively this means that there are few external incentives for breeders to go the extra mile for their breeding stock. It is also true that those with atrocious levels of care and hygiene are rarely penalized.

With quick and dirty dollars to be made, a backyard breeder can upgrade to a much larger scale operation and become what we call a puppy farmer, a truly detestable character that makes Cruella de Vil of *101 Dalmatians* fame look like Mother Teresa. The emergence of puppy farms or puppy mills focusing on the production of numerous pups of poor to mediocre quality to be dispatched all over the world has been disastrous for puppy welfare. If puppies spend the first weeks of their lives in a kennel so filthy that they never learn the difference between clean areas and dirty areas, they will inevitably be difficult to house-train.

With their emphasis on numbers of pups, greedy breeders aim for two litters a year from each bitch. Immature and older bitches often struggle to regain bodyweight as a result of having to produce so much milk. And don't be misled: Puppy farms are often the sources not just of pedigree dogs but also of so-called designer crosses. The recent demand for labradoodles and then spoodles, schnoodles and cavoodles was met with enthusiastic, grubby hand-rubbing by puppy farmers everywhere.

Another grave concern about puppy farms is that they rarely bother to socialize their pups and so tend to produce companion dogs that are fearful of everyday stimuli. Given that this leads to fear and ultimately manifests as aggression, you can see why inadequate socialization is one of the main reasons people regret buying a pup that was born on a puppy farm.

Pet shops are often affiliated with puppy farms and are criticized for inadequate out-of-hours supervision of animals, poor socialization of pups and for using pens that discourage house-training. The shops are often designed to prompt impulse purchases of young animals by people who are ill-equipped to care for them and may also facilitate animal teasing by members of the public. Most welfare groups deplore pet shops for these reasons and, of course, because they add to the total number of unwanted pets and increase the workload for shelters.

Inherited disorders

Ever since I was a veterinary student, I've been perplexed by the problems posed by inherited disorders. At the University of Bristol my undergraduate colleagues and I were constantly being reminded that diseases A, B and C were common in dogs of breeds X, Y and Z. Indeed, the situation began to seem overwhelming. Remembering which of the 180 or so breeds had which of the 400 or so diseases was challenging enough, without beginning to design a system to tackle the issue. To some people, allowing animals to be born with a preventable disease is a form of cruelty.

It is only fair to point out that dogs are not alone in having unwelcome traits imposed on them. Indeed, when you discover that some pigeons are bred specifically to tumble instead of fly—a life-threatening trait—you see how far animal welfare can rank behind human gratification. More commonplace pets, such as cats, pocket pets and caged birds, all have their own particular lists of inherited disorders, but dogs take the lead. They offer the best example of the many problems that can arise when pedigrees and show-ring success are prioritized over health, temperament and welfare.

Before the first dog show, dogs were judged on their ability to work for us as hunters, retrievers, herders, guards and so on, or their ability to gratify us as companions. When dogs left the working arena for the show-ring in the late nineteenth century, many of the functional aspects of their behavior and morphology were incorporated into breed standards. The intention was noble—to provide a record of the ideal representative of the breed—but the language used was flowery, convoluted and open to misinterpretation. In fairness, at the time photography was not readily available, and writing was the most appropriate means of creating such a record. The results of judges' interpretations of the written standards are the pedigree dogs we recognize today.

There was a belief that if the animal is able to work in its original role, its conformation (body structure) must be absolutely correct. Hence, the mantra "form follows function"—meaning if it's put together the right way, it's bound to work. Unfortunately, this sort of short-term approach to dog health and welfare is way too simplistic. Looking as though you are able to do the work of your ancestors is no guarantee of good quality of life. Also, reliance on written breed standards risks a lot being lost in translation. As we will see in this chapter, standards can be contradictory, counterproductive and open to interpretation. Often seen as having been handed down on virtual tablets of stone, they are guarded with religious fervor by traditional breeders, who refuse to accept that the written word can leave much to the imagination and resist any changes to the standards even when dog health is clearly being compromised.

Unfortunately, some of the breed standards exalted in the show-ring place more importance on appearance than on functionality. Effectively, they are all show. Breeders compete among themselves to work out who can produce the phenotypes that best conform to the written breed standard. Longevity and trainability—two traits that companion dog owners value—are not scored in the show-ring and are therefore not selected in breeding plans. In practice, many of the breed standards include traits that have, at best, questionable welfare and health benefits. For example, the Saint Bernards currently winning in the show-ring may meet the breed standards, but, plagued by hip and heart problems, this breed is a lumbering, slobbering health crisis with an average life expectancy of less than five years. Any stranded alpine adventurer hoping to see one coming round the mountain with a full barrel of five-star brandy is just as likely to see Saint Bernard himself.

Aural health

With even a superficial glance at dogs of various breeds we can see how body structure and health are related. Just as head shapes and eye shapes differ from one breed to the next, so, too, do ear shapes, and this can have a significant impact on health and welfare.

Hanging ears are extremely appealing, especially in pups, but sadly they predispose the animal to problems caused by poor ventilation. An unbalanced population of microflora, either fungal or bacterial, can proliferate in the warm, moist haven of such ears. The wax and pus that accumulates in the ears of middle-aged spaniels is something to behold. As a vet who regularly made house calls, I was able to tell the moment I entered a living room whether it was home to any aging cockers: their itchy ears create a telltale tidemark of aural filth against the walls and furniture. Pendant ears may also trap grass seeds, another perennial problem for spaniels. Unfortunately, these foreign bodies are difficult to flush out, and the dog must often undergo general anesthesia to allow the vet to use an auriscope and forceps. Hanging ears also droop into food bowls, unless you feed the dog from a high-sided container that pushes the ear flaps away from the food.

Pricked ears, on the other hand, seem to be a target for biting flies, perhaps because they are obvious, in contrast to folded ears, which may be hard to spot against the background of fur on the neck. They also seem more prone to the risk of being deformed by blood blisters (aural hematomas are really only a particular problem for dogs with these ears) and, as a result, developing into cauliflower ears, the trademark of a human boxer.

Hair, especially in the ears of dogs bred for a fine silky coat (such as Maltese, poodles and Yorkies), readily blocks fresh air from the ear passage; so hairy ears can be prone to the same problems as hanging ears. They need to be plucked, usually under general anesthesia.

Bred for the show-ring but not for life

Let me illustrate my point with a number of examples taken from current Australian breed standards, which align with the global dog-breeding body, the Fédération Cynologique Internationale (FCI).

I have selected a number of the more stylized breeds to show how the standards may run counter to good welfare and how easily extreme characteristics specified in breed standards can be linked to disease.

Let's start with Weimaraners, whose breed standard demands that the chest is "well developed, deep" while the abdomen is "firmly held" and the flank "moderately tucked-up." These requirements may help to make Weimaraners appear athletic but vets know that breeds with deep chests are at risk of gastric dilation and torsion, an extraordinarily painful, life-threatening condition in which the stomach bloats with gas and can become twisted. Then we have the pug that, according to its breed standard, should have eyes that are "very large, globular in shape." Breeders oblige the judges and breed for this feature. But is it then just a coincidence that pugs have eyes that bulge so badly their lids scarcely meet well enough to wipe the eyeball clean? The poor dogs undergo a lifetime of chronic conjunctivitis that eventually scars over the cornea and effectively blinds them. They are well and truly "puggered" for life. Meanwhile, breeders of British bulldogs, are told that the "head should be very large—the larger the better." Favoring big heads and narrow pelvises in the show-ring means that pups (with big heads) get stuck in their mother's (narrow) birth canal. So this is the breed most likely to need a caesarean section to survive birth. This same breed is also required to have curved "roach" backs. Unsurprisingly, they are sometimes born with twisted spines.

Sometimes breed standards are blurred, contradictory and confusing. For example, the Japanese chin's head "should be large in proportion to the size of the dog, broad skull, rounded in front and between the ears but never domed." It is tricky to see how a skull can be broad, rounded in the front and between the ears, without being domed. Similarly, the shar pei must have "loose skin" and a "frowning expression," but these characteristics should "in no way disturb the function of the eyeball or lid" and dogs should be "free from entropion" (an unpleasant condition in which the eyelids are rolled in towards the eyeball). The truth is, this combination of loose skin and frowning expression must predispose dogs to entropion. The problems do not end with the eyes, though, since the wrinkled skin of this breed can, like any sweaty crevice, become raw, rubbed and smelly. I'll die happy if this book reduces the naïveté of even one potential puppy-owner who thinks shar peis are so cute that all they'll ever need is just a big hug. Staggering but true: Some pet shops sell shar peis with a complimentary voucher for the plastic surgery the pups will need as they mature. This sort of approach to dog welfare makes me grind my teeth in frustration.

Another example of the gaping gulf between the show-ring and reality is the puli's standard, which describes a temperament that is "wary of strangers." The desired appearance of this Rastafarian-looking favorite is to have "long hair [that] overshadows [the] eyes like an umbrella." Simply cutting the dreadlocks obstructing the dog's vision gives it more warning of approaching humans and so immediately reduces its need to be wary. Beyond that, why would anyone seeking a companion animal want one from a long line of dogs selected, not least, because they were wary of strangers?

In some cases, traits that would compromise Feral Cheryl's chances of survival are incorporated into breed standards. For example, shortness of the skull (brachiocephaly) is described and generated by the standard for Boston terriers that require animals to be "short-headed" and to possess a "short head and jaw" with a muzzle that "is short, square, wide and deep . . . shorter in length than in depth; not exceeding in length approximately one-third of the length of the skull." We know that short skulls are associated with breathing problems. Indeed, many middle-aged Boston terriers, pugs, cavaliers and French bulldogs need surgical trimming of the soft palate to open their airways.

Even the Australian native dog is under threat, with dog fanciers

showing interest in having dingo shows. For many observers this
speaks of either considerable naïveté or arrogance. How can we
improve on animals that have succeeded in a hostile environment for
many thousands of years? What qualifies anyone to write the breed
standard for such animals?

Baby face—the selection for neoteny

As dogs made a transition from working to companion and show
animals, lap breeds were selected for their puppy-like features. An
example, with its *large dark round eyes, pendant ears and compact cush-
ioned feet*. is the cavalier King Charles spaniel. And so the perpetual
puppy has found its evolutionary niche in the caring nature of some
humans. As those doe eyes softened our hearts, this selection for
pedomorphism may have strengthened the human–canine bond.
Unfortunately, it also reduced the visual signaling capacity of many
breeds (e.g., the French bulldog with "rose" ears that cannot be
pricked up). Generally speaking, the welfare consequences of a
reduced signaling capability have yet to be established, but although
unreported and unexplored, it seems likely that it may make socializing difficult for some breeds.

Let's pause here to consider how dogs perceive departures from
the basic dog shape. I'm not suggesting that dogs may regard poodles as members of a strange sect, but even non-puppy-like breeds
probably present a challenge for observing dogs. Some breeds, such
as the Maltese, are not so much too neotenized as simply too hirsute
to signal effectively. Their long-haired guinea pig good looks lead
many dogs in parks all over the world to mistake them for prey. As a
further example, whether or not they can wag their (bob)tails, Old
English sheepdogs have difficulty raising their hackles (because their
hair is too soft and long), displaying bared teeth (because of their
beards) and delivering fixed stares (because of their veils).

From a behavioral viewpoint, a number of features of pups are worth noting. They generally have a relatively high tolerance for unfamiliar humans and dogs; a dependence for food, care and leadership; a readiness for play; and an acceptance of substitute objects to elicit hunting behavior. These essentially juvenile behavioral traits have all been favored by breeding. One possible explanation for this is that modern companion dogs have been selected to exhibit a behavioral repertoire that is primarily passive, deferent and accommodating. This makes them less assertive than Uncle Wolf would ever have been at maturity and makes them virtually dependent on the human group for many of their activities. Conveniently, it also helps them bond with us and value us as feeders, caregivers, coaches and super-heroes. Despite this, we keep them on their own. In Australia, for example, 75 percent of dogs live in single-dog households. And then we wonder why they bark when left alone. We have made them dependent only to routinely leave them in isolation.

All of this prompts us to consider the environments in which most companion dogs spend their lives. Although pedigree dogs are generally bred from successful show dogs, most of them end up as companions in our homes, an environment that requires a very different set of characteristics than the show-ring. And the hard work for which many breeds were originally intended has been replaced by restricted exercise in inner cities. Many modern dogs also spend much of their lives away from their owners. A cynic would say that we should really be breeding dogs with innately lower exercise demands and a reduced predisposition to separation-related distress. However, selecting against separation-related distress could mean selecting for reduced dependence or even heightened self-confidence, which may lead to resource-guarding and dog bites.

Problems with current breeding practices

Breeders sometimes devote more energy to refining the quality and color of their dogs' coat than to caring for the health of the wearers. As with horses and cattle, selecting for color has resulted in some unforeseen and unwelcome changes in families of dogs. For example, there appears to be an association between coat color and aggression in cocker spaniels. Self-colored (or solid-colored) cockers (those without white hairs in their coats), especially golden ones, are more likely to glaze over and enter a state of so-called rage than roan-colored cockers, for example.

Behavior patterns favored in a working context can sometimes be inadvertently over-selected and give rise to compulsive tendencies. Border collies have been selected to "show eye" (stare), and many now demonstrate a fixed stare at blank walls. Perhaps it is the rat-killing legacy in staffies and bull terriers that makes some chase their tails and chew them to bits if they catch them. The debate continues as to whether animals that show such obsessions are actively suffering when they perform their repetitive behaviors. Some argue that the repetitions may actually be a way of getting high on endorphins. While this so-called self-narcotization may mask the true extent of a welfare problem, such obsessive compulsive disorders (stereotypies in nonhuman animals) can interfere with normal behavior and, in extreme cases, prompt euthanasia if the owner finds the behavior unacceptable.

The wrong priorities?

With every generation in a breeding program you can make only a certain number of strides. Since breeders have to take into account the many detailed traits specified in breed standards, there's limited room to apply added pressure to breed for traits that relate to welfare and adaptability to urban environments. Strong selection could be imposed for relevant temperament and performance but only if less attention were paid to traits of peripheral importance.

Performance in this instance doesn't simply refer to athletic performance. The minimum performance required of a dog is the ability to survive birth without assistance. In contrast, genes (such as those in the huge-headed bulldog pup attempting to emerge from the wasp-waisted mother) that can be passed from one generation to the next only with the intervention of a vet are inherently faulty. It can be argued that both dam and offspring have failed an essential performance test. Unfortunately, market forces are not effective in discouraging caesarean births. On the contrary, it is common for both breeder and veterinarian to benefit financially from this practice, because the cost of surgery is passed on to purchasers of the pups.

One option for veterinarians faced with the ethical dilemma of operating to save the lives of oversized pups *in utero* is to perform an elective pan-hysterectomy at the time of the caesarean. While ending the bitch's breeding career, this approach does little to terminate the genes partly responsible for the oversize, since the bitch's female pups will carry a similar tendency.

Dangerously shallow gene pools

Pedigree studbooks are closed. This means that all breeding stock have to be recognized as members of the breed and have family trees filled exclusively with other registered members of the breed. Rather like the old rules for breeding royal humans, no new genetic material is allowed in. This means that any strides made within a generation in a pedigree dog-breeding program are, at best, only baby steps. Even without the pressure to conform to breed standards, many breeders would still find themselves producing dogs with serious defects, since almost every animal that has ever lived has carried at least one harmful recessive gene. The average number of harmful recessive genes carried by an individual dog can be as high as 20. Disorders caused by harmful genes have been listed in various reviews, books and databases. An on-line catalogue, Online Mendelian Inheritance in Animals, OMIA (www.angis.org.au/Databases/BIRX/omia) includes more than 495 inherited disorders reported in dogs.

Even when disorders are not life-threatening, they may still cause significant welfare problems. These can range from the frustration of being less able to play due to respiratory problems, in the case of brachiocephalics (short-skulled dogs) with compromised airways, to the distress of corrective surgery in dogs with orthopedic problems. Unsurprisingly, novel defects continue to be reported (for example, "twitchiness" in miniature wire-haired dachshunds), and some of these will certainly turn out to be inherited.

While natural selection limits the frequency of harmful genes, the mating of relatives (inbreeding) increases their impact dramatically. Inbreeding brings harmful recessive genes out into the open, where their effects can be seen. The two take-home messages are clear:

- Even the purest of purebred animals is likely to be carrying harmful genes.
- The greater the level of inbreeding, the greater the chance of producing dogs with inherited defects.

Although dog breeders should aim to avoid mating close relatives, this is almost impossible in numerically small breeds because it is often extremely difficult to find a mating pair that does not share ancestors within a couple of generations. Surprisingly, popular breeds with very large numbers of registered animals are not exempt from this problem, either. The actual rate of inbreeding can be far higher than one would expect from the number of dogs registered because too many breeders concentrate on just a small number of families (often called line-breeding, which is just another word for inbreeding). It may give the illusion of short-term benefits, such as being able to say that a litter is sired by a very popular stud, the world-renowned, Supreme Champion Flavor of the Month, but the long-term price is that other bloodlines are lost, and the choice becomes ever more limited. Think of inbreeding as peeing in your own gene pool.

The challenges

Good breeders want to see these problems addressed. With a long-term perspective they can appreciate how elusive significant change will be under the present system of competing against written breed standards alone and banning out-crosses (the introduction of new bloodlines). Essentially, they are doing the best they can under the current rules but generally agree that pedigree dog-breeding faces at least five major problems:

1. Some breed standards and selection practices run counter to the welfare interests of dogs, to the extent that some breeds are characterized by traits indefensible on welfare grounds.

2. There isn't enough pressure to select traits that would improve animal welfare and produce dogs better suited to modern human living.

3. The incidence of certain inherited defects in some breeds is unacceptably high.

4. The number of registered animals of certain breeds within particular countries is so low as to make it almost impossible for breeders to avoid the mating of close relatives.

5. There may be financial disincentives for veterinarians to reduce the incidence of inherited disease.

Steps in the right direction

The status quo is that most breeds have their characteristic list of inherited defects, some of which occur at an unacceptably high frequency. Of course, there is money to be made by treating these disorders, so there could be financial disincentives for veterinarians to help control them. Sadly, this was confirmed when a senior veterinary colleague asked me to stop highlighting the issue of inherited disorders since it was the bread and butter of the profession. Happily, he has since retired, and I am heartened by the fact that more than 250 Australian veterinary practice managers have agreed to report inherited disorders on a regular basis via the Listing of Inherited Disorders in Animals (LIDA, www.vetsci.usyd.edu.au/lida, a website hosted by the Faculty of Veterinary Science at the University of Sydney). This is a giant step in the right direction, since it allows us to know our enemy. If we know which disorders are the most common in the pedigree dog populations of each country, we can develop strategic breeding plans to minimize their prevalence. You can find out more about inherited disorders in dogs by visiting LIDA, our database of inherited disorders that allows you to search for a specific breed group and organ system.

The central aim of the LIDA project is to collect and process the data that will allow interested parties, including breeders and veterinarians, to monitor the prevalence of inherited disorders. We are doing so by:

- developing software that monitors certain fields in veterinary practice management databases currently used in Australian practices
- collecting these data centrally and collating them to identify the most numerous disorders per breed and the age at which they most commonly present to veterinarians

- sharing the information with veterinarians, breeders and potential puppy purchasers at no charge.

Another of our initiatives is to identify a way to encourage selection of breeding dogs based partly on temperament. The central idea is an award for show dogs that have passed a standardized temperament test. It has received strong support from the Australian Small Animal Veterinary Association (ASAVA) and Delta Society, and in-principle support from the Australian Veterinary Association (AVA), Australian Companion Animal Council (ACAC) and Australian National Kennel Council (ANKC).

The award will be presented at the major shows in each capital city. Dogs that have passed the test will be invited to nominate for this award. Following the normal judging of dogs entered in the show, the highest-scoring dog in each group (toys, gundogs, terriers and so on) that has nominated for this award will win the award.

We acknowledge that breeders could train a dog with a questionable temperament to squeeze past the test. However, it is hoped that breeders will be encouraged to select easy-to-train temperaments. This, after all, is what our companion dogs need to have.

CHOICE CUTS

- 🐾 There are important differences between what a breeder and a veterinarian would consider to be a high-quality litter.

- 🐾 The world of dog-breeding is full of unscrupulous characters keen to make easy money at the expense of animal welfare.

- 🐾 Beware of pet shops that are affiliated with puppy farms. Animal welfare at such breeding grounds leaves a lot to be desired, and the pups sold are often hard to train.

- 🐾 An unacceptably high number of inherited defects occur in some breeds.

- 🐾 Some of the breed standards prioritized in the show-ring place more importance on appearance than on a dog's temperament and well-being.

- 🐾 Some breed standards and selection practices run counter to good welfare in dogs; in fact, many incorporate traits that would compromise a dog's chances of survival in the wild.

- 🐾 There should be more pressure on breeders to select for traits that improve an animal's wellbeing.

- 🐾 Breeders should be encouraged to breed dogs better suited to modern environments.

- 🐾 Closed studbooks make gene pools dangerously shallow.

Dingoes on Fraser Island are said to be the purest genetically, since they have interbred only minimally with domestic dogs. Their general behavior, social order and interactions with humans offer important lessons for dog watchers. It is regrettable that some people, ignorant of the dangers of some current dog-breeding practices, want to exhibit such dingoes in competition, using a written breed standard.

17
Rex in the City

We may look forward to a time when dogs are selected for their ability to adapt to modern living, but this is easier said than done. A major obstacle here is that the goalposts keep shifting. The urban niche dogs occupy has changed tremendously over the past 50 years and it continues to do so. Looking at past and current trends that affect dogs can give us a whiff of the shape of things to come and where we are sending the dogs of the future.

The dogs of yore

In the bad old days, dogs were rarely restrained and so would regularly wander the streets, scavenging in bins, stealing from hostile shopkeepers, having sex and dodging vehicles—pretty much the way they still behave in developing countries. Owners might have taken their dogs with them to the corner shop, or not. It very much depended on the dogs themselves. From a dog-welfare perspective, this was far from ideal. Some got lost, some got squashed and others were stolen, but many successfully organized day-to-day routines and daily routes, nodding to the officer on his beat, greeting the granny in her park and worshipping the butcher surrounded by sacrifices in his sawdusted temple. All manner of neighboring doghood would pack together and patrol their patch. So, their daily exercise and their territory were very different from what we see in most cities today.

Mothers largely stayed at home, which meant that they could provide almost constant company for the household dog when he wasn't on patrol. Perhaps this explains why separation-related distress was less of a problem then. And don't forget, the dog next-door was probably not barking, either. This is significant. Not only are dogs likely to bark in response to others barking, they may even become distressed by the noise and thus become more likely to develop separation-related problems of their own. In the quieter days of black-and-white movies, the compounding effect of a barking dog on neighboring dogs was minimal.

Spending on the dogs of yesteryear was limited to the purchase of a collar from the general store, the occasional bone from the butcher's sawdusty shrine and a newspaper to be rolled-up tight for random acts of mindless discipline. The original dog food was simply human leftovers. The science of dog nutrition has since blossomed so that now there are cheap foods, premium foods, canned foods, dry foods, semi-dried foods, semi-moist foods and specific diets for specific disorders. Apart from price, the major issue is the relative palatability of these products versus that delicious and potentially unhealthy mound that had just been scraped off Grandad's plate. Pet food manufacturers learned very early on to urge us to avoid feeding scraps to dogs, arguing that balanced nutrition formulated by veterinarians and laboratories is safer than scraps can ever be.

Modern dogs

Generally, the percentage of owned dogs in developed and developing nations is increasing and with it the number of dog-related goods and services. Whereas they were originally needed to help us work—with hunting, herding and guarding—dogs in the modern setting help us cope with the stresses of modern life.

Smaller families and longer working weeks mean that modern dogs are not only forced to spend more time on their own but also may be less experienced than their predecessors at interacting with both familiar and unfamiliar people. Modern dog owners must control their dogs, which means confining them to the home property for most, if not all, of the time and certainly when owners can't supervise them. At the same time, higher density living tends to bring dogs and their owners into closer contact with neighbors, who may or may not own dogs themselves. All this means that dogs need to be more tolerant of both confinement and of strangers.

Dogs behaving as we want them to give us enormous pleasure, but the reverse is also true. Stress relievers can sometimes be stress creators. We probably all like to imagine that the pros outweigh the cons, but how can we be sure? Very few of us can claim to know whether our dog's behavior is average or above. We don't have any common standards since there's a lack of communication between owners, and, of course, many of us would rather not face the ugly truth.

We already know that we put up with a great deal of canine misbehavior in our homes. This was confirmed by a recent survey of UK dog owners showing that the full litany of unwanted behaviors is as long as a designer dog's shopping list. When asked to evaluate their dog's behavior, 51.8 percent of owners reported their dog had an annoying habit (mainly barking) and 25.3 percent described at least one behavior (usually either aggression or disobedience) as a problem. The good news? This was offset, at least in part, by the 64.5

percent who described their dogs as having endearing habits. Our tolerance of unwelcome behaviors reflects the role dogs play in our lives. After all, if dogs are supposed to help us cope with stress, there is little point in getting stressed by the animals themselves.

Nowadays, pet food is the biggest-selling range for most supermarkets. The aisle devoted to snacks for dogs is one of the fastest-growing areas in supermarket real estate and is flanked by the emerging wave of pet warehouses and centers for doggy day-care. This speaks not only of an increase in disposable income but also has a whiff of owner guilt. We spend to show we love.

Canine internal medicine has also advanced in enthusiastic leaps and bounds, so while dearly beloved James Herriot was once considered sophisticated for wielding a stethoscope, modern vets are making diagnoses based on ultrasonography, gamma scintigraphy, CT and MRI scanning.

As citizens of developed nations, we undoubtedly benefit from a tremendous array of technological advances, but we are also subject to relentless social, cultural, economic, political and technical change. The pace of change in modern lives is matched only by the rate of flux in modern relationships. Interactions, intimate and otherwise, between adult humans can be stressful, reflecting the high expectations we have of one another and our relationships. Stress and its management have become central to our lives as we increasingly sense isolation, a loss of control, and sometimes even loss of confidence in the future. Gee, it's scary! Against this backdrop, owning a dog can be therapeutic, because the stability it represents can help us relax and be ourselves. Importantly, these benefits accrue within a reliable relationship. As such, dogs form part of what has been called the back-to-basics agenda. People walking their dogs in a city park are mimicking the agrarian lifestyle of our forebears, temporarily rejecting technology and the need to produce, consume and communicate.

Modern dog ownership offers several significant challenges, not the least being that the relationship between dogs and the community in general has altered. Many people no longer own or interact with dogs regularly. Mothers grab small children as soon as they see a dog. The child is taught to fear dogs and learns to scream as if being boiled in oil when a dog approaches. Accordingly, dogs are segregated by town planners. So, despite the importance of adequate socialization in preparing dogs to cope with life in a human community, there are fewer opportunities for them to socialize.

Leashes are far less optional than they used to be—nowadays they are a sure sign that owners value their dogs and take responsibility for them. It's interesting to note, though, that to some observers, leashes are a sign of oppression. Some animal rights activists argue that dogs should not be fettered in this way but should be free to make their own choices in life. These choices presumably include the option of being mashed by a truck when they choose to step off the pavement at the wrong time. For those who wish to keep dogs safe from traffic, leashes are the only way to go.

For anyone wanting to guide a dog as humanely as possible, the choke chain is unacceptable. Over the past 20 years, head collars such as the Halti and Gentle Leader have emerged to replace not only choke chains but also regular flat collars made of leather or nylon. The principle, borrowed from the world of horse training, is simple: Where the head goes, the body will follow. These head collars are so effective it is difficult to see why they were not developed for dogs when they were developed for horses, cattle and camels. Certainly, creating a plain collar is simpler, and perhaps dogs were just that much easier to pull around by the neck than, say, a horse, which would be more likely to bolt, taking the would-be leader with him.

City living

As apartments become more common, many bodies corporate are allowing residents to keep a dog. This is good for owners and, in some respects, also for the dogs themselves. The health benefits include a reduction in disease transmission and exposure to parasites, soil-based fungi and garden poisons. And certainly dogs confined to apartments are not going to be hit by cars. It is even possible to hope that they get a better view of the world around them than the average garden-dwelling dog and, at the same time, may be spared the barrier frustrations of many a backyard—there will be no children rattling sticks along the fence, for a start. The negative health consequences relate to confinement and apply as much to any confined dog as they do to those living in apartments. Balconies on some apartments can accommodate a dog's need to urinate. However, limiting a dog's opportunities to urinate increases the risk of bladder stones, while few opportunities to defecate effectively may result in constipation. The recent invention of pet toilets for apartments is of concern if the devices allow owners to imagine they don't need to exercise their dogs so much. Dogs on farms that have few structured walks but virtually unlimited freedom to dump wherever and whenever, space four to six droppings throughout the day. This is the natural system for removing ingested food from the bowel once it has had the goodness removed. Having to retain feces for extended periods is not good for either digestion or temper.

Confinement is a serious issue, but the size of an enclosure is less important than its contents. Bland, barren, boring environments are far worse than small yet entertaining ones. This is the principle of environmental enrichment that is driving exhaustive efforts in zoos, farms, stables and laboratories to improve animal welfare by providing for the needs of each species. Chapter 3 explored the value systems of dogs and emphasized the importance of making resources relevant to the ethology of the dog. So, getting it right for dog ownership in the city depends on meeting the dog's behavioral needs, not on having a backyard.

The drawbacks of dog ownership

Dogs can even be victims of their own success. They can cause distress just by being so well loved. For example, the anguish caused by losing a dog, or even simply imagining such a loss, can make some owners hesitant to replace them.

Generally speaking, dog ownership is not necessarily either successful or easy. Following are some of the reasons that dog–human relationships are pear-shaped.

- Dogs tie you down.
- Dogs cost money—not so much the purchase price and food costs, but the ongoing and unexpected maintenance costs, such as vet fees.
- Sometimes dogs can be dirty and smelly.
- Dogs can cause friction with neighbors.
- People may not have dogs that are appropriate for where they live.
- Some people are allergic to dog fur and saliva.
- A dog may fail to fulfill its owner's expectations.

The last of these issues seems to be the most monumental. I blame Lassie for raising our expectations way beyond reality. These days, dogs in movies, with the inexhaustible assistance of Inspector Rex, continue to promote an unrealistic perspective on what a good dog can do for us. Educational initiatives that help children to let dogs be dogs are a perfect response to this.

All the drawbacks listed above can result in dumping and surrendering dogs to pounds, but they could all be avoided with good planning. That said, they do not reliably deter future ownership because they are countered by the benefits.

Dogs versus cats

I think it's worthwhile to step back and ponder the merits of dogs versus cats. It's been said that dogs are from Mars and cats are from Venus. Both cat and dog owners agree that the two species differ so much in behavior that the nature of the relationship one has with them is also bound to be quite different. Dogs are generally thought to need more care and effort than cats, but they are also considered to be more emotionally accessible and more likely to sustain a rewarding relationship. As well as being easier to train, dogs are considered, by both cat owners and dog owners, more likely to show need, love and gratitude than cats. However, research by Maree McCallum, of Sydney, has shown that cat owners refute dog owners' claims that dogs are more faithful, clever and steady. It seems that cat owners value behavioral traits that speak of discrimination and taste. Of course, for non-cat people these are exactly the traits that seem like arrogance and disdain.

In dog-loving countries, announcing that you are not a dog person can imply that there is something wrong with you rather than with dogs. This is not the case with cats. Cats polarize attitudes more than dogs. In Australia, especially, people are more likely to have a prejudice against cats because of the perception that they wreak havoc on native wildlife.

Prejudice and some sexual stereotyping also influence many people's perceptions of dogs versus cats. For example, cats tend to have female connotations, whereas dogs are often perceived as macho. Similarly, the sensuality and mysteriousness of *Felidae* intrigue some cat lovers, but the same attributes are sometimes regarded as devious and sneaky by dog lovers. To the uninitiated and uncommitted, cats can seem unpredictable, reserved and self-centered. Many cat owners concede that cats may not always be as approachable and friendly as dogs but argue that this is because cats are more independent and discriminating, less simpering and ingratiating. When cats tend to bestow their affection, recipients can feel privileged and somehow anointed. By contrast, dogs can be something of a pushover.

Generally speaking, cats can require more emotional work even though they may be cheaper to feed and are usually less demanding. With their more outgoing approach to life, dogs are easier to get on with and so more immediately rewarding. While they may need more care and be more dependent, it is this very dependency that makes owners feel needed and loved. Dogs also get people off their butts and so can bring health benefits way beyond the reduced blood pressure that comes from stroking anything furry. Dog ownership forces us to be more active. An example of this phenomenon is Walk the Dog, a recent national advertising campaign launched jointly by the RSPCA (Australia) and the National Heart Foundation.

The benefits of modern companion dogs

There are many good reasons not to own a dog, *so why do we bother?* Ultimately, we have to admit that the logic of pet keeping is elusive, and it is chiefly the emotional, rather than the practical, benefits that motivate us. Like falling in love or having children, owning a dog is a zone we enter despite being aware of the problems and hard work entailed. Yet, however irrational, having a dog clearly brings many benefits.

Due to the pressures of contemporary life, more people now want to own pets. Modern relationships are often the product of family fission and fusion. The notion of jobs for life is as obsolete as choke chains, and this can mean a long-term lack of stability and confidence. It's been suggested that, as human relationships become more complex, pets provide a welcome contrast. So, as families morph, businesses merge, streetscapes stretch to accommodate high-rise developments, dogs can represent a constant, a cornerstone in a household. Regardless of how testy all the humans in the household are, the dog can usually be relied on to make the best of things.

The routines, structure and rituals of dog care, walking, feeding, watering and grooming can give owners tremendous purpose and counteract feelings of loneliness and isolation. There is evidence that concentrations of oxytocin rise in owners when they merely look into their dogs' eyes and that exposure to dogs can make adolescents more emotionally competent. Whereas decorum and moderation dictate our behavior with humans, we express ourselves more freely with animals. Dogs can be cuddled, chatted to and played with, so they can be receptacles and catalysts for emotional outpourings. Unlike some people, dogs generally respond to contact. Even people who insist they are "not from a hugging family" and are physically standoffish with humans often seem to show a great need to cuddle their dogs.

Dogs can even facilitate human contact as they inform and enrich family life and sometimes represent neutral territory during family disagreements, disputes and divorces. The returns dogs give on a low emotional investment can be massive, especially when compared with some human relationships. Although relationships with dogs don't require the same input of time and effort as those with humans, rewards in the form of gratitude, fun and affection generally flow thick and fast. Additionally, dogs keep on giving, even when the owner has been transiently absent or neglectful.

Dogs can take us back to basics. People who are "good with animals" have an undoubted social cachet and are often great conversationalists, as long as the topic is animal-based. When permitted, dog owners relish the opportunity to talk about their animals. Informal permission for this sort of conversation is granted freely by other pet owners far more frequently than for indulgent conversations about one's children. Scientific studies show that dogs provide a reason to talk to strangers, and so dogs in parks increase social capital. Pet stories are also a safer staple of journalism than sex, religion and politics.

Since dogs cannot answer back or complain, owners assume a leadership role, and this undoubtedly gives many of them a degree of satisfaction. In a sense, they can play God. They "giveth and taketh" and even use their pets as pawns in their relationships with other humans. It's possible that the power owners can exercise over their dogs may even compensate them for their perceived lack of power in relationships with friends, family and work colleagues. This may be borne out by the undercurrents of competition one sometimes detects between the two owners of a shared dog as they vie to demonstrate the control they have over it. But I believe that, on the whole, power over dogs is seldom abused, and the suggestion that dog owners are control freaks simply came from a leaflet dropped by a low-flying cat-owner.

The roles of modern companion dogs

The future role of a newly acquired dog can lie anywhere between being a furry slave and a hairy child. Despite that, and regardless of the intensity of their feelings for their dogs, most owners still regard dogs as animals, not humans, which is why euthanasia, for example, is an accepted practice for those that are suffering. Whether we like it or not (and often whether the dogs like it or not), dogs can be playmates and animated, malleable toys, especially for children. But for many of us, the value of dogs is chiefly to do with their pure appeal as animals. Lovers of animals in general don't treat their dogs as fur-coated humans but are fascinated by all that is quintessentially canine about dogs.

Clearly, dogs play many different roles in their owners' lives. For example, they can be useful when teaching children about tricky topics such as death and reproductive behavior. Indeed, parents of younger children often nominate this as a reason for having a pet. Some parents even see keeping a pet as part of their responsibility as parents. Typically, if they had pets as children, parents regard it as a right for their children to have them too, and expect the pets to educate the children to be humane, respectful and compassionate. Although reality rarely fits these expectations, children often bargain for a pet with their parents by agreeing to care for and feed it.

A dog's role in a household differs with its age and, indeed, the owner's age. Dogs in their dotage—whose ears, skin or breath have become putrid with neglect—are often cruelly banished from the house. Fortunately, the reverse is also true: A favorite working dog may become a housedog, a ratter from the yard may slither in on a cold winter night and manage to avoid expulsion forever after, and racing greyhounds may—if extremely lucky—change gear to become couch potatoes. A puppy bought for children may become an (aging) child substitute when the human children leave home.

The role of pets as child substitutes has probably been overstated. The image of Tricky-woo, the flatulent Pekingese we were told about by James Herriot, was a caricature. I offer this judgment because, as a vet in general practice for more than five years, I hardly ever encountered dogs in this role. In contrast, the requests from my friends for Tricky-woo-type tales were relentless. The assumption is that anthropomorphism peaks in childless dog owners, who use the language of a parent whenever addressing their dogs. I reckon such comments were usually made in jest, only to be leapt upon by pop-psychologists.

The myth of the modern dog as a furry kid is reinforced when puppy buyers recount how they had to prove their worthiness to own a particular puppy. Clients of mine have commented that being interrogated by breeders of a certain pup was more rigorous than getting a home loan. However, given how rare it is to have unsold pups in the kennels of well-organized and reputable breeders, I wonder if the clients are more anxious to let me know that they passed the examination than the breeders ever really were in listening to their answers.

Dogs can be status symbols, fashion accessories or just one of the trimmings perceived as necessary for the perfect family. Absurd as it seems to many of us, breeds go in and out of fashion. A great example is the Afghan, whose popularity in the sixties was truly phenomenal—there were even specialist vets who practiced in nothing but Afghans. Nowadays, Afghans are chiefly show dogs.

Whether as a fashion statement, embellishment or distraction, a breed can help to define the identity of the owner, because it reflects their personality and interests. Pit bull terriers barge instantly to mind. Their ability to pack on muscle makes them a perfect prop for any pinhead with a petite penis. Some even have to wear body-building harnesses and share steroids. The macho motivation for muscular male dogs means that bitches are barely bothered with and older dogs are dumped. Dogs of any breed can assume a function in affirming the owner's authority because they are often more amenable to discipline than other household members and rarely challenge the owner. In many cases they can be easier to train than children, and in some cases they may even help adults understand the importance of consistency in educating their children.

Urban animal management

Urban animal management is an emerging field that attempts to reconcile society's need for companion animals with the problems of urban living. On the face of it, most companion animal legislation is driven by a need to control defecation and procreation. But other problems include dog attacks, excessive barking (a leading cause of community disputes) and, mainly in the case of stray dogs, road accidents, livestock losses and property damage. The central thrust of urban animal management is to promote responsible pet ownership.

EDUCATION VERSUS LEGISLATION

The therapeutic benefits of pet ownership are so important to owners that they tend to mistrust any interference in their private relationship with their pet. On the other hand, most animal lovers want to see negligence towards pets punished and responsibility or duty of care firmly assigned to owners. This is where urban animal management legislation comes into play. Urban animal legislation is designed to force owners to take more responsibility for their dog's behavior in public but also to protect the rights of both dog owners and the rest of the community while still allowing owners to enjoy the rewards of dog ownership.

Clearly, it is more palatable to limit the number of puppies being born by neutering than it is to cull adult animals, and legislation that encourages neutering, rather than insists on it, is rarely controversial. Similarly, people tend to welcome legislation that protects wildlife from domestic pets, even though there are many practical difficulties in restraining cats, and some cat lovers argue that night-time curfews for what is essentially a nocturnal hunter verge on cruelty. The flip side is that legislation tends to affect dogs more than cats, and there is generally more community support for dogs than cats. So, in principle, dog owners may support legislation that addresses these issues even though they may be emotionally resistant to it.

REGISTRATION

Legislators ask that dogs be registered to force owners to take responsibility, a pivotal requirement when communities attempt to manage dangerous dogs. That said, the owners of a stray dog that has caused a road traffic accident could be as reluctant to come forward as the owners of a dog that has bitten a child. Registration policies are usually designed to address this problem; the main snag here is that the very people who are less diligent about restraining and training their dogs are also least likely to register them.

A local or national registry should help to:

- reunite lost pets with their owners
- identify the owners of problem animals (such as dangerous dogs)
- monitor local trends in companion animal ownership
- collect pet-owning fees.

Registration relies on accurate and easy identification. Identity tags on collars can be lost, and they make it necessary for the pet to wear a collar, which some owners see as a strangulation hazard. For these reasons and because they are virtually tamper-proof, subcutaneous microchips have become the favored means of identifying companion animals.

CONTROL

Responsible pet owners keep their animals under control to prevent them from causing a nuisance and endangering themselves. For a dog, being under control means being:

- effectively contained on the owner's property
- when off the owner's property, held on a leash by a responsible adult.

To some extent this limits the exercise dogs can take of their own volition, but it has the undoubted benefit of reducing the risk of road-related trauma to companion animals.

Dangerous dogs

Responsible pet owners recognize that some dogs can be hazardous to humans and to other animals. Depending on the region, local or national legislation may govern what becomes of dogs that are labeled dangerous. Sanctions on dogs that have been declared dangerous range from increased control requirements through to compulsory neutering or to compulsory euthanasia.

Dangerous dogs can be identified by either "deed or breed," that is, what the dog has done in the past or because of its specific breed. Of these two approaches, the latter is highly contentious, since it places restrictions on dogs simply because they look like one of the fighting breeds. In many cases the most dangerous thing about a pit bull terrier is the owner at the end of its leash. Crossbred dogs that happen to look like members of one of the restricted breeds can be impounded in error and become the focus of exhaustive legal action. Some such dogs have spent years in custody while their breed and therefore their fate was decided. The welfare implications are cumbersome and complicated but include the issue of confinement, possibly for extended periods. The discussion of temperament tests in chapter 14 may help you decide whether you support early detection of dogs that may be a menace to society.

Networking among dog owners

Although we are generally meeting our dogs' need for food, we might not be meeting their need for fun and exercise. While urban dog-keeping may be becoming more of a grind, there are technological solutions that can grease the wheels. Networking among dog owners can help us to get our dogs tired. Information technology has been used to put humans in touch with one another, and now perhaps it's time for our dogs to hook-up, too (see, for example, www.dogtree .com.au).

In the future

The growing importance of dogs in modern living seems to be borne out by the clear trend towards premium pet care. In developed countries, a great deal of effort, time and money are being spent on pets in general, and the booming pet-care industry and merchandising veterinarians are part of the response to this trend. This is not to say that dog owners in bygone days did not care about their animals as much as we do, just that things were different then. For example, flea products were relatively useless, so dogs were less welcome in a house, let alone on a lap; vets were largely trained as physicians for horses and production animals, and pet food meant dog biscuits. These days, many dogs sleep on their owner's bed, and some lines of dog food outsell all products intended for humans. There are so many small animal vets that they are selected on the basis of their branding and bedside manner (and reputations for being specialists, even in the diseases of a single breed).

Where are these trends going? Will vets compete too fiercely for clientele? Will they over-service and end up cooking the goose that currently lays double-yoked golden eggs? Will pet food manufacturers work out how to package bones, grass and other animals' feces? Will the claims made for pet food plateau and purchasers come to regard them all as much of a muchness? Will double beds simply have to get bigger?

With time, stem cell therapy may also emerge, most likely riding on the back of human advances in related disorders. For example, it is hoped that genetic intervention into human diabetes might also be harnessed to assist diabetic dogs. And, of course, cloning is now a reality. It will be fascinating to see how clones are received into the homes of families who loved the previous user of the DNA enough to want to replace it. The extent to which a newly arrived clone behaves as its predecessor once did will expose the critical significance of learning and training by design, stealth and accident.

Legislation seems likely to increase rather than diminish. The non-dog-owning public will get used to having certain expectations, and the emotional, working and heritage value of the dog may well be overlooked. Increasingly, we are encouraged to complain, and we live in a consumer-driven society that has allowed us to expect satisfaction or at the very least action. This may be part of why there are more registered complaints about barking dogs. People are more likely to complain when unruly dogs romp through a park and send picnickers reeling or, worse, knock over a small child. In the eyes of the law, harm done by dogs is increasingly likely to include fear. So, even in the absence of physical injury, dogs that have frightened children may, in certain instances, be declared dangerous.

As the consequences of dog-owning become better recognized, we will gather more information on both the good and bad news associated with dogs, and the ideal family dog will become more recognized and celebrated. As we evolve together, my hope is that the "loving, half-comprehending, half-mystified aliens who live within our homes" will increasingly become less mystifying and alien, while remaining just as loving.

Conclusions

As we become more aware of the physical and emotional needs of pets, pet care becomes less easy but possibly more rewarding. Overall, humans these days live better, and, in many cases, so do dogs, especially those that score the best life-coaches.

The apparent liberty and bond with its owner may make a street dog's life seem idyllic, but the downside is the danger of traffic accidents, which means that such dogs rarely make old bones.

The benefits of bringing children and safe dogs together under direct supervision are numerous. While acquiring empathy, many children learn about responsibility and care for others through formative relationships with dogs. Children who have poor dog skills and show fear are unfortunately likely to trigger predatory aggression.

Acknowledgments

Ben, Nessie, Annie, Wally, Tinker, Neville (all of whom are pictured in this book), and all the other dogs in my life have been my greatest teachers and companions; I thank them for forcing me to take time in recreation and appreciation of the natural world. They have shown me the value of dog games, consistency in training and social order in groups of dogs, but these lessons are eclipsed by demonstrations of the dog's extraordinary behavioral flexibility. In their ability to adapt to a multiplicity of niches, dogs are unsurpassed. How regrettable it is that this is ever taken for granted or abused. Dogs deserve our love, respect and admiration. I hope their ability to overlook our carelessness, ignore our behavioral inconsistencies and forgive our social clumsiness will one day be fully appreciated.

In my university teaching, I try to use a research-led approach that has been informed by canine studies from my own research group. These include those conducted with Tanya Grassi, Tristan Starr, Alison Harman, Alex Brueckner, Abby Masters, Lara Batt, Hannah Salvin, Lisa Tomkins and Taryn Roberts. Teaching questioning undergraduate students further enriches my work and develops my approaches to dog behavior and management. I thank the University of Sydney's Faculty of Veterinary Science for the wonderful teaching and research opportunities it has presented to me. Among my colleagues at the University of Sydney, I am particularly indebted to Emeritus Professor Bob Boakes, who has helped my understanding of learning theory.

Fellow dog fanatics who have provided invaluable editorial advice for this book include Pierre Malou, Stephen Ryan, John Miller, Jason Johnston, Ruth Mackay, Mark Robertson, Emma Lawrence, Anne Stubbs and Karin Bridge. Nick Branson and Mia Cobb gave particularly welcome expert advice on chapter 5.

Further assistance was provided by my mother, Mary McGreevy, who originally saw dogs as the archenemy of home hygiene but, when faced with an adorable gift dog, finally relented to her children's persistent requests for a dog. Mary, thank you for making such an excellent decision and allowing the passion of two of your brood to blossom and lead them into the veterinary profession.

Norm Keast and Marie Rowe were generous in granting permission to use photographs. I also warmly thank Stephen Pincock from UNSW Press and Nadine Davidoff for their ideas, enthusiasm and encouragement.

My great friend and mentor Lynn Cole has yet again been most generous with her time in editing various drafts.

Index

About the Author

PAUL MCGREEVY is an Associate Professor at the University of Sydney's Faculty of Veterinary Science and author of several books on animal behavior. His pioneering work has attracted many awards, including the British Society for Animal Science RSPCA Award for Innovative Developments in Animal Welfare, the Australian College of Veterinary Scientists Ian Clunies Ross Memorial Award and Universities Federation for Animal Welfare Companion Animal Welfare Award.